Zero to Culture

Zero to Culture

A STEP BY STEP GUIDE TO IMPLEMENTING AN EMPLOYEE ORIENTED SAFETY MANAGEMENT SYSTEM

First Edition

Ron Dotson

Eastern Kentucky University

SAN DIEGO

Bassim Hamadeh, CEO and Publisher
Angela Schultz, Senior Field Acquisitions Editor
Michelle Piehl, Senior Project Editor
Alia Bales, Production Editor
Jess Estrella, Senior Graphic Designer
Stephanie Kohl, Licensing Coordinator
Natalie Piccotti, Senior Marketing Manager
Kassie Graves, Director of Marketing
Jamie Giganti, Director of Academic Publishing

cognella® | CUSTOM
3970 Sorrento Valley Blvd., Ste. 500, San Diego, CA 92121

Contents

CHAPTER ONE

How to Lead Safety Culture

FOREWORD

Leadership is the key to success in any endeavor. Safety professionals of many titles in diverse settings must realize that influence is their key factor for determining success. Leadership deals with people and motivating them to exhibit leadership traits in their behaviors. While leadership is not a title of authority, and exists at all levels of authority in any organization, we can say that the goal of safety professionals is to influence safety at the individual decision level, treating it as a virtue rather than a priority that changes situationally. It is vital that values be developed that influence others at all levels because new hires will seek out mentors. This chapter will examine the theory behind the application and serve as a guide to the application of leadership in establishing a culture of occupational safety as concurrent with the goal of activity or production.

This chapter will define safety culture through moral decision making; connect decision making to leadership; examine scholarly studies; and, finally, tie leadership practices and measuring culture to safety.

Objectives
After reading this chapter the learner will be able to:

- Describe safety in the context of culture
- Determine the difference between safety as a priority and as a virtue
- Examine and recognize the criteria of cultural levels
- Relate Theory X and Theory Y assumptions to safety policy
- Recognize the base for empowerment in safety decisions
- Construct the key elements of leadership
- Assess organizational ethics
- Formulate methods to measure ethical commitment and safety culture
- Organize leadership practices to protect the integrity of group decisions

LEARNING PLAN

LEVEL	ULTIMATE OUTCOME	PROBLEM	LEARNING TYPE	ASSESSMENT
U	Describe safety in the context of culture Determine the difference between safety as a priority and as a virtue Examine and recognize the criteria of cultural levels	The goal of safety leadership is to establish safety as a virtue in decision making.	CA/CT	Students will present on safety leadership to include: defining leadership, goal of safety leadership, virtue versus priority, influencing associates at the management level and workforce-associate level, and committee/problem solving leadership techniques, working in examples of cultural levels.
U	Relate Theory X and Theory Y assumptions to safety policy Recognize the basis for empowerment in safety decisions	Safety managers must be capable of predicting policy outcomes based on assumptions made about those participating in the process, as well as provide the resources necessary for actual empowerment in decision-making processes.	CA/CT	Students will develop an overall management philosophy that reflects Theory Y assumptions and empowers workforce-level associates in safety-related decision making, while concurrently holding participants accountable for their performance.
U	Construct the key elements of leadership Assess organizational ethics	Managing organizational conflict requires the establishment of organizational values and a commitment to adhere to these values in making organizationally ethical decisions.	CA/CT	Students will develop a leadership section for an overall safety management plan that sets vision, defines and exemplifies core values and guiding principles, and defines measures for organizational ethics.

U	Formulate methods to measure ethical commitment and safety culture	Safety managers must establish standard measures for the organization that track performance of safety as a subset of culture in order to demonstrate successful safety performance.	CT	Students will establish specific measures for safety performance based upon the cultural criteria of: participation, commitment, attitudes and perceptions, competency, and compliance.
U	Organize leadership practices to protect the integrity of group decisions	Groupthink phenomenon is detrimental to organizational ethics.	CT	Students will establish committee positions and duties in order to prevent groupthink during safety committee processes.

Learning Plan Legend

Level:

F: <u>Foundational Outcomes</u>: basic abilities

M: <u>Mediating Outcomes</u>: Progress through a developmental model; interpret, analyze, evaluate progressively challenging claims, arguments

U: <u>Ultimate Outcome</u>: Navigate most advanced arguments/claims

Type of Learning:

CR: <u>Critical Reading</u>: The ability to read, process, and understand the meaning of written information

IL: <u>Information Literacy</u>: Locating and selecting suitable information for a task; evaluating appropriateness/validity of information sources

AC: <u>Application of Concepts</u>: Ability to apply discipline-specific knowledge/skill to tasks/situations important to the discipline

CT: <u>Critical Thinking</u>: Ability to apply a concept to a vague or argumentative claim without a creative leap.

AT: <u>Analytical Thinking</u>: Ability to critique/analyze situations using a concept or model.

CA: <u>Creative Application</u>: Ability to apply a model/concept in a new way to an unrelated situation or scenario. Involves creative leaps.

THE MORAL ARCHITECTURE OF SAFETY

Culture is commonly known as the values of an organization. Whether it is a family, a country, a group, or a company, the values common to all or prevalent among the members are displayed in actions. Wagner and Simpson draw a distinction between morals and ethics that in the past did not exist. Morals are personal to a member of the organization and then permeate the moral architecture of any organization of which they are a member (2009). Decisions are made based upon personal morals. These decisions play out in the behaviors exhibited by any organization. Ed Schein teaches us that assessment of organizational culture is three tiered; at the surface is the symbolism or visible courtesies, traditions, stories; the middle level consists of policy and procedure; while the root level consists of the organizational values (1988). A culture of safety in an organization then becomes a matter of leadership making the connection between personal morals that form the root values; reflecting and developing those in policy, procedure, and decision making; and finally establishing visible safety practice in the organization's symbolism. Safe conduct must therefore be moral and a personal virtue.

The moral architecture of an organization has several elements that impact a vision of safe conduct. According to Wagner and Simpson's model of moral decision making, courtesies, ethical principles, virtues, democratic processes, personal relations, policies, attitudes, goals, commitments, habits, and communication patterns all influence the organization's decision making as priorities (2009). The degree to which these criteria are held to or brushed aside in decision making reflects the moral architecture. It might perhaps be a better model if virtues were culled out of the circle of morals and placed outside of and in direct influence of the others. Virtues are the character of any person. The higher degree of a virtue, the more the person resists minimizing the virtue in the making of any decision. People of higher virtue rely less on rules to govern their actions (Zagzebski, 2004). Therefore the goal of safety leadership is to instill safety as a virtue that is unchanging from situation to situation, as is the case with a priority.

People make decisions with priorities in mind that change situationally. However, many people have core virtues that do not change from situation to situation. These virtues are never violated. These can change from person to person, but those personal virtues are unyielding.

Figure 1.1 reflects the moral architecture model suggested by Wagner and Simpson with character virtues influencing the moral architecture that defines the safety culture of an organization.

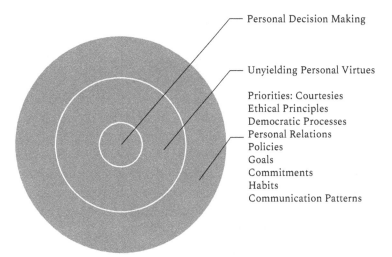

Personal Decision Making

Unyielding Personal Virtues

Priorities: Courtesies
Ethical Principles
Democratic Processes
Personal Relations
Policies
Goals
Commitments
Habits
Communication Patterns

FIGURE 1.1 Moral Architecture Model

Virtue theory began with Aristotle (Wagner & Simpson, 2009). Virtues were an average between two extremes that people migrated toward based on their experienced feedback. The virtues were courage, honor, patience, loyalty, self-discipline, empathy, honesty, and compassion (Wagner & Simpson, 2009). Virtues develop over time based upon the positive and negative occurrences, or feedback, associated with them. For example, bravery was considered to be a point midway between rashness and cowardice. The person would be more or less brave depending on their positive and negative perception of times when they acted either on the rash side or on the cowardly side. They would migrate somewhere in between (Ciulla, 2003).

Virtues are important influencers on the moral architecture of an organization. Virtues are critical traits that define leadership. Leadership influences the culture. Therefore, decision making reveals the culture. In terms of safety, when an employee makes a decision to work at a hurried pace, that decision has moved some of the elements of moral architecture to the rear and brought other elements to the forefront. In other words, the value of finishing quickly has replaced the motivation to maintain a safe workplace. Safety management then must establish initiatives or methods within a safety management system that keep desired safety behaviors a priority. Roughton and Mercurio describe the goal of a safety management system as reducing the amount of negative risk that the members of the organization are willing to accept (2002).

Management must display virtue in the moral architecture; the policies, the commitments, the principles, the communication patterns, the goals, the habits, the democratic processes, and their courtesies. Only then can the safety culture reflect safe conduct as a value.

Silent consent is one phenomenon that reduces safety from a value to one of many priorities that can be juggled to the front and rear. When a person of authority in an organization fails to correct the acceptance of negative risk, it establishes a rejection of safety for some other priority, thereby making safety a priority rather than a virtue. To ignore or condone an unsafe act or condition is to grant permission to exempt safe practice for production or quality. Numerous practical examples exist, including managers who ignore spills; fail to fix machine guarding safety controls; bypass safety protocols; fail to properly equip personnel; or pretend not to see a clear safety violation, such as not wearing proper personal protective equipment. Such activity becomes silent consent because it grants permission to dismiss safety in light of the drive to produce for profit. As a friend of mine and production manager used to say, "Safety is first, but production is king!"

In the long run, safety must be a value. Even a shortcut or near miss without any injury or property damage produces a cost. Any negative occurrence at least contributes to future incidents that may have direct costs. So production is not taking place if it accepts negative risk. Deming brought quality to the forefront of production culture by establishing the formula for production as Product + Quality = Production (Aquayo, 1990). But the formula must be Product + Quality + Safety = Production.

All members of the organization must accept that working safely is not a right but rather a fundamental value, just as honesty is a virtue. Safety is not a right that workers are fighting for against management. It is already their own value. It cannot be prioritized or exercised when it is convenient. Safety cannot be allowed to be a dividing point between labor and management.

Empowering safety begins with adequate training and education on recognizing hazards and abatement strategies. It continues with initiatives that grant responsibility for identifying and correcting hazards. Objective measures of performance reinforce positive actions. Authority to stop work and enforce safe conduct solidify the value. It is then that safety will appear at the lowest level of safety culture—visibility. Desired behaviors, visible signage, absence of obvious hazards, traditions, stories, and other symbols will make safety apparent.

Safety empowerment is never granted if one does not have the authority to stop the work in order to correct an unsafe condition or act, or to at least consult with safety and management about the concern.

The elements of moral architecture must promote virtues. Guiding values for making safe decisions must be set and established. These guidelines allow for all associates to make decisions that are free from negative repercussions if the results are less than favorable. Without guidelines, associates at any level will be hesitant, at best, to make decisions. Decisions made and justified as consistent with guiding values are good-faith decisions. Organizationally, these good-faith decisions are ethical.

The moral architecture suggested by Wagner and Simpson and adapted to reflect safety as our central theory of betterment lists the following as criteria for culture:

- Courtesies
- Principles
- Democratic processes
- Personal relations
- Policies
- Goals
- Commitments
- Habits
- Communication patterns (2009).

Schein's model for viewing organizational culture poses three levels. The root level is a collection of the values. The values in turn influence the policy and procedure of the organization. Finally, this plays out at the immediate level. This is the visible spectrum. The visible spectrum consists of the behaviors, traditions, and other symbols observable in the organization (1988). Schein's levels of culture give us a general outline for assessing safety culture. The model of moral architecture makes the model more specific. One can correlate the models as in Table 1.1 below:

TABLE 1.1 *Model of Moral Architecture*

CULTURAL LEVEL	CULTURAL CRITERIA
Root Level	Virtue/Guiding Values
	Principles
Mid-Level	Goals
	Commitments
	Policy
	Personal Relations
	Communication Patterns
	Decision Making/Democratic Processes
Immediate Level	Courtesies
	Stories
	Traditions
	Visible Symbols
	Behavior Patterns/Habits

The key to culture is to develop the values into the immediate level. The mid-level must actually encourage and develop the immediate level of culture. This process begins with the works of Douglas McGregor on the human aspect of organizational management. He identified two opposing views of human nature in regard to managing the human resource; Theory X and Theory Y. McGregor teaches us that in the making of any decision or the taking of any action, management makes assumptions about human behavior. Theory X has three assumptions:

1. The average human avoids work if possible,
2. Because of the tendency to not work, most people must be coerced into putting forth acceptable effort, and
3. The average person wants to be directed, thus avoiding responsibility, and desires only security.

Management policy and actions influenced by these assumptions or promote these assumptions are Theory X type actions (2006).

Theory Y is in stark contrast to X. Theory Y actions assume or promote the following assumptions:

1. The average human puts forth efforts toward work as naturally as play or rest. Satisfaction can be gained through work.
2. Humans will exercise self-direction and control in the furtherance of objectives with which they agree.
3. Commitment to objectives is directly proportional to the rewards associated with their achievement.
4. Avoiding responsibility, lack of ambition, and desire for security are only learned from negative feedback and are not natural.
5. The capacity to creatively solve problems is widely distributed in the population rather than narrowly distributed.
6. The intellect of the average human is only partially used and must be promoted (McGregor, 2006).

In short, an organization that is Theory X in safety-management philosophy blames workers for incidents without addressing management-system factors, concentrates on punishment for violations rather than correction, incentivizes bottom-line performance indicators, and relies on rules to be enforced. By contrast, the company that is Theory Y in safety management philosophy places safety as a value and does not prioritize safety, addresses management-system factors in root cause, recognizes positive behavior, and empowers all associates with safety responsibility. Safety is viewed not as an enforcer but as an equal subpart to the other subparts of the organization, a subpart that relies on participation in order to function and deliver services to the other subparts.

McGregor compared human resource management to engineering's approach to specification problems or engineering challenges. He compared an engineer's efforts to control problems by staying within the natural laws of physics. An engineer would not design drainage to flow against gravity, for example. His point is that management efforts toward human resource, or control, often fail to consider the assumptions and proper motivation for human action, in effect violating human nature. McGregor theorized that society would tolerate management's control efforts to meet objectives only so far as it protected and preserved human values (McGregor, 2006).

McGregor suggests that underlying the efforts to control were the virtues of leadership characteristics. These values trumped and limited management efforts. This is why his work on assumption became so important. Managers were making underlying assumptions that were being made about human behavior at work. He suggested that identifying underlying assumptions before management action was key to anticipating behavioral response. For example, regarding incentive plans, McGregor points out that the underlying assumption is that people want money and will work harder to get it. Therefore management measures the job task and sets a production scale with pay that is a bonus for above-standard production. He does point out, however, that the plan for control fails to recognize other assumptions. As far as incentive plans are concerned, management fails to realize that

> "most people also want the approval of their fellow workers and that, if necessary, they forgo increased pay to obtain this approval; (2) that no managerial assurances can persuade workers that incentive rates will remain inviolate regardless of how much they produce; (3) that the ingenuity of the average worker is sufficient to outwit any system of controls devised by management (McGregor, 2006, pg. 12)."

In relating Theory X and Theory Y to safety we can make some general assumptions. Theory X may assume workers do not want to work safely if it requires additional effort and that due to additional efforts to be safe, people must be coerced to follow safety procedure and that the average human prefers to be coerced toward safety, perhaps to satisfy an ambition to overcome fear. In sharp contrast, Theory Y may assume that expending effort toward safety is as natural as efforts for work, play, or rest and that workers, if committed, will self-direct efforts to be safe. In addition, Theory Y assumes that satisfaction of ego and self-actualization can be achieved with safety objectives toward the individual worker and production teams, that resistance to safety is learned and not inherent, that workers are capable of critical and creative application toward solving safety problems, and that there is untapped potential for safety innovation in the average worker.

While experienced safety managers may report that they have experienced or witnessed both X and Y assumption in worker behavior and may postulate that you could apply the concepts situationally, McGregor would counter that management actions influence behavior. In short, if Y type behaviors are desirable, actions must originate in agreement with Theory Y assumptions. The safety policy and procedures must encourage the development of Y type behaviors (McGrgeor, 2006). People will rise to the level to which they are treated, so to speak.

McGregor was a Theory Y proponent. One striking observation he made may shed light on lackluster safety performance in spite of tight controls. It involves active resistance to management control efforts. Once workers have their basic needs fulfilled, Theory X assumptions in action do not allow the worker to grow in meeting higher-level needs, such as self-actualization. When this occurs, workers may rebel against management controls. Since safety is a basic need, right after physiological needs on Maslow's scale, as shown in Figure 1.2, safety must encourage self-actualization through actions consistent with Theory Y.

TABLE 1.2 *Safety Assumptions*

THEORY X	THEORY Y
• Workers do not want to work safely if it requires additional effort. • Due to additional efforts to be safe, people must be coerced to follow safety procedure. • The average human prefers to be coerced toward safety, perhaps to satisfy ambition to overcome fear.	• Workers will make the effort to work safely if safety resources are present. Resistance to safety efforts are learned, not inherent. • Workers will self-direct efforts to be safe • Ego and self-actualization can be satisfied through meeting safety objectives • Workers are capable of solving safety problems and have untapped potential.

Maslow's Hierarchy of Needs

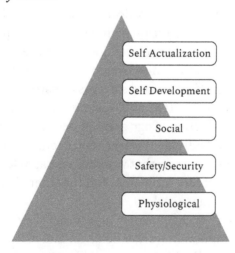

FIGURE 1.2   Maslow, Abraham. (1943). A Theory of Human Motivation. Psychological Review, vol. 50, pp. 370–396.

Empowerment is undoubtedly a scary concept for true Theory X type managers to accept. Many may say they believe in worker empowerment, but true empowerment is only achievable by leaders who have a high degree of self-confidence in their own abilities and faith in the talent of their people (Maxwell, 1993). In regard to safety, empowerment begins with the most basic investigational skill, hazard recognition. Workers must have a model for recognizing hazards and then progress to a model for correcting those hazards. This begins with a psychological approach to safety, education, and training. The next step would be to develop that talent through a formal, systematic program for hazard recognition that truly relies on worker action. The safety manager takes on a role of confirming program findings or producing a measure of accuracy and then of overall program performance, or reporting on the program's performance and how it correlates to safety objectives, plus encouraging frequency and accuracy of worker involvement. This goes without saying because workers cannot self-actualize safety if they cannot recognize and develop basic countermeasures, or hazard controls, at the level of their trained competency.

Workers do not approach work by finding hazards first, as a safety professional would. Humans recognize fears or concerns after they begin work activity. This is the natural path for humans, otherwise phobias might limit humans' ability to meet their own basic needs. In 2012, Sandseter and Kinnear published a study on the children's play, identifying six levels of risky play. These included, "exploring heights, exploring speed, exploring dangerous tools, exploring dangerous elements, wandering alone, and rough and tumble play" (Sandseter & Kinnear, 2012, p. 265). The hypothesis that resulted from the play was that children play in order to overcome fears. This supplemented earlier theories of play, such as the theory that children play to develop physical skills needed for survival. But the background for the exploration of fears originated from a type of treatment for adults with phobias. The treatment involved controlled exposure and exploration of the fear along with individually paced overcoming or abatement of the fear (Sandseter & Kinnear, 2012).

It makes perfect sense that humans explore fear early during play. Otherwise, we might develop phobias that inhibit necessary action for productive activities that expose us to risk (Sandseter & Kinnear, 2012). In other words, our ability to assess risk and make reasonable judgments as to the acceptance of risk would be impaired. It is undeniable that overcoming fears is a necessary quality for many human actions of value. A good example is exposure to dangerous elements in order to save another's life.

Our first step toward empowerment, which can result in egotistical satisfaction and self-actualization in safety, is to teach workers to recognize categories of hazards that they encounter. In teaching categories, our goal is to identify the categories in a way that provides easy recognition without too much overlap. These categories are:

Impact

Penetration

Compression

Chemical

Respiratory

Temperature

Visibility

Radiation

Walking/working surface

Electricity

Animal, insect, vermin

Biological

Noise

Ergonomic

The next step in safety empowerment is to develop abatement skill, first by covering basic abatement hierarchy and then by developing methods for managing change and solving problems. If safety is established as a virtue that is not to be prioritized but rather adhered to unconditionally, then the worker is capable and ready to self-direct safety.

REFLECTION 1.1

1. What is the difference between a priority and a virtue?
2. Does theory X management practice contribute to safety being a priority of convenience?
3. Can workers be expected to make proper safety decisions in a theory X environment?
4. Where does the empowerment to be safe begin?

THE LEADERSHIP CONNECTION

Confident leaders allow empowerment. Empowerment is a higher level of participation and is indicative of effective leaders. Typically, leaders who are transformational are more effective, partly at least because participation increases job satisfaction (Zainnuddin & Ibrahim, 2010).

Leadership is a process of influence. The leader motivates, inspires, and influences others' behaviors. But the subordinate allows or receives the influence. Leadership then exists at all levels of an organization. Leadership is not by position. Members of the organization that have authority to enforce rules are managers by title, not leaders. Leaders exist among numerous subsets of the organization. This includes groups of functional workers as much as certain groups of friends. These leaders are the organization's key advocates, or those that are listened to.

Leadership's role in any organization is to influence in a positive manner. Therefore, the safety manager must influence all associates and develop systematic methods for key advocates to positively influence associates in regard to safety. Leadership involves communication by coaching, motivating, directing, guiding, and supporting others (Howard, 2005). Influence relies on traits that in turn guide behaviors of the leader.

Effective leaders have four characteristics: direction, trust, action, and hope (Howard, 2005). Overall direction is provided collaboratively by establishing the means for guiding values and building virtue in the symbolism of the organization. Secondly, leaders establish trust through consistent adherence to values. Others can predict accurately the attitude and behavior of the leader. Leaders are action-oriented and not afraid to sacrifice for the good of the team. Hope is also conveyed through empowering others to participate.

Research in opinion leadership indicates that strong personality traits that facilitate closer ties to others is more of a predictor of influencing opinion than specific knowledge of the issue. While knowledge is important, personality traits and word-of-mouth communications are much stronger in influencing one's opinion, attitude, and overt behavior (Gnambs & Batinic, 2013).

McCrae and Costa define personality traits as "dimensions of individual differences in tendencies to show consistent patterns of thoughts, feelings, and actions (1990, p. 23). We are taught that leadership traits include self-discipline, courage, commitment, initiative, honesty, integrity, tact, bearing, honor, knowledge, humility, compassion, loyalty, respect, patience, and a sense of justice (Dotson, 2017, p. 180). Research reveals that these traits are consistent between people, are tied to heredity, and influence behavior across diverse situations (Kornør & Nordvik, 2004). Traits play a primary role in the behavior of leaders.

Foundationally, there are four types of leaders. Type A leaders are concerned about facts. They conservatively analyze data—more quantitative data than qualitative—and are very task oriented, or give details and close oversight. Type B are creative leaders. They are artistic, flexible to other views, and act spontaneously. Type C leaders are emotional and make decisions based on personal views. They are more extroverted, intuitive, and rely on interpersonal relations. Type D leaders are control based and rely on strict structure. They are sequential learners and planners; competitive; formal; and very task oriented, likely to over-direct tasks with even experienced persons (Howard, 2005). All of us probably have varying degrees of all types based on our personality traits. What matters is learning to adjust and use our own strengths in ways that accomplish positive influence.

Behaviors and styles depend on our raw traits. Traits are developed through our own experiences. Just as Aristotle believed virtues were means of behavior, impacted by the positive and negative experiences that individuals had over time, we use or display these traits to the left or right of the mean. We learn to have courage somewhere between being too rash and too cowardly depending on our own positive and negative feedback when we acted more cowardly or more rashly (Ciulla, 2003). Traits influence style and behaviors (Kornør & Nordvik, 2004). Style and behaviors develop into dimensions of behavior and of traits in academic research.

Overall behavioral dimensions are classified as charismatic, transactional, transformational, and visionary (Kornør & Nordvik, 2004). Charismatic leadership is influencing action and enthusiasm in others through personal attributes and behaviors (Zehir, Muceldili, Altindage, Sehitoglu, & Zehir, 2014). Charismatic leaders are draw on your emotions to influence. Transactional leadership is when leaders emerge from deals and compromises. Leaders position themselves to bargain and compromise with subgroups in order to gain more favor with each group. These leaders rely heavily on theory X type rewards and punishment in order to ensure results (Ruggieri & Abbate, 2013). Transformational leadership involves follower participation and support (Ruggieri & Abbate, 2013). The transformational leadership model involves servant type leadership, where the leader serves the follower in a way that furthers the organization's objectives while facilitating the growth of the follower. Visionary leaders provides a vision of the future, turns it into objectives, and shares these with others. The visionary leader must convince others that the new direction is a common interest. Charismatic leaders differ from visionary in that they use emotions and charm to influence others in the direction of their own personal views, which may not be a vision for the good of the organization.

Research points toward three dimensions of leader behavior; task-oriented behavior, relations-oriented behavior, and change development. These dimensions are predictors of influence (Kornør & Nordvik, 2004). Leaders who are more effective in task production are not easygoing, open, or self-confident leaders. Instead, they tend to resemble Type D leaders. Leaders who are more suited for employee relations exhibit trust, straightforwardness, altruism, compliance, modesty, and ease. Leaders suited for change situations are more emotional, and they are set on values, ideas, and actions. Leaders who are suited for situations of relations and change tend to resemble Type B and C leaders (Kornør & Nordvik, 2004).

Team building and morale is a major concern for safety managers and is more effective in transformational practices. Participation and consultation result in higher job satisfaction (Ismail, Zainnuddin, & Ibrahim, 2010). Leader self-sacrifice builds team identification (Ruggieri & Abbate, 2013). Leader characteristics suited for team building are creativity, being flexible and open to input, and being passionate, intuitive, and extroverted. These leaders build and rely on relations. Transformational leaders use coaching, develop subordinates, and empower others. This allows self-actualization for the subordinates.

Team building is an important aspect of leadership. Research shows that organizations oriented toward team dynamics, such as team goals, work groups/committees, and that use group-participation methodologies have higher performance levels and higher job satisfaction than organizations oriented toward individual behavior (Hinsz & Nickell, 2004).

Leadership serves as the foundation for establishing safety as a virtue. Leaders cannot influence without building trust and exemplifying the actions they desire from others. Transformational strategies build morale and team affiliation. It begins with character traits and virtues. Maxwell's "Law of the Lid" predicts team success based on the totality of team leadership potential (Maxwell, 1998, p.1). It also begins with self-development of traits and virtues. Of primary importance is the study of leadership and leader behaviors in order to temper personality strengths and build weaknesses. Safety leadership must then build the leadership potential of associates in their everyday policies and procedures. The goal is empowerment as the next step up from participation.

REFLECTION 1.2

1. Where does transformational leadership begin? Why?
2. What are some key ways you would build team identity?
3. Why is hope a characteristic of a leader?

MEASURING CULTURE

Authority creates the environment of two cultures within an organization when it comes to safety. The culture of safety can differ between workforce and management personnel. Here we can define management as frontline supervisors and their authoritative superiors in regard to authority, or chain of command. The workforce then comprises the associates who are not supervisors who schedule or direct activity. Typically, subordinates in the workforce mirror the culture of safety displayed by management. But this is not always a certainty. Safety is only one aspect of the overall organizational culture, and other aspects of culture permeate all subsets of the overall culture. Therefore, safety management must develop culture for management and for the workforce. This necessitates measuring safety culture from both perspectives.

Culture begins with organizational ethics. Organizational ethics refers to how well group decisions reflect organizational values. Organizational values are set and emphasized in order to minimize conflict. Ethics begin with virtue, personally translated to group ethics as the group determines or the leader sets guiding values. Virtues are commonly listed as courage, patience, honor, loyalty, respect, discipline, honesty, and compassion. Virtues of safety originate from these in order to guide all associates toward ethical safety behaviors. Table 1.3 below lists the safety virtues.

TABLE 1.3 *Safety Virtues*

COURAGE	To do the task safely To report hazards To correct hazards To correct undesired behavior
PATIENCE	Not to take a shortcut To consult with task experts To consult with others
HONOR	To hold yourself and others to working safely even when no one is watching
LOYALTY	To the safety of your fellow associate
RESPECT	For the well-being of self and others
DISCIPLINE	To not prioritize safety
HONESTY	In safety participation
COMPASSION	For new hires and returning casualties

To measure an organization's ethical decisions, gauge how well its behaviors match its values. Several criteria can be examined in order to measure organizational ethics. These should include:

- Do policies reflect organizational values?
- Do policies develop safety virtues?
- Does management make decisions consistent with organizational values?
- Are the values displayed in the symbolism of the organization?
- Are organizational values reviewed?
- Do all associates have a chance to provide input about the review of values?
- How does the workforce perceive management's adherence to values?

The challenge is to find systematic and objective measures in answering the criteria for ethical assessment. One practical procedure would be to assess whether the decision or project coming out of committee meets goals and is consistent with organizational values. Any group decision, or strategic or practical application can be assessed by the group. Standard committee practices should include going over guiding values after establishing group consensus of the problem in order to guide abatement strategy. One group member should also be assigned to assess group decisions and discussions based on organizational values.

It is commonly recognized that culture in an industrial organization can actually have two coexisting subcultures. One subculture is the culture of the workforce associate, while the other is the culture of the management-level associate. The workforce consists of the rank-and-file workers who are assigned schedules and roles. The management level begins with frontline supervision and extends upward in a hierarchical structure. Safety management system standards recognize these subcultures and refer to them as management commitment and employee participation. Each culture can have five measures that define it: participation, commitment, attitudes and perceptions, competency, and compliance.

Participation means to take part in something. Management participation in safety can appear in many forms and must be considered from the frontline supervisor role to executive levels. We begin measuring management participation by examining whether management level positions are allocated in cross-functional committees that safety participates in. Active participation means more than a sign-off role where management is informed of committee actions and progress. We must also examine whether management allocates safety representation on policy-making committees to include strategic planning and business-strategy committees. Many organizations allocate safety representation on safety committees, disciplinary committees, problem-solving groups, and similar decision making groups, but fail to include safety in bid processes, business continuation, strategic analysis, or committees seen as only production related. Management participation must be the involvement of management personnel in safety-related committees or groups as well as management-allocating positions on business-related groups for safety representation.

Specific ideas for measures include the percentage of relevant committees on which safety has an allocated position, as well as the percentage of safety-related committees on which management has a position of representation at the supervisor and manager levels. Attendance rates are another measure for management participation. Having an allocated position does not mean that attendance occurs.

Other measures of management participation depend on the core management programs that safety will use to deliver services to its various customers. For example, if a behavior-based safety program is in place, a requirement for observation or other form of participation by supervisory and management personnel would provide a program measure for management participation. Participation components present in all safety management initiatives can also be summed as a management participation measure.

Management commitment is altogether another measure that defines a safety culture. The ultimate role of upper-level management is to provide the necessary resources that enable safety to meet its goals and mission. There is one final measure that indicates whether this has been done by management; the ratio of closed countermeasure reports in proportion to open requests. In other words, after the safety department has identified the hazard and problems, produced a countermeasure, and presented it for approval, it is up to the next level of management to allocate resources. When these resources are allocated, management has supported the safety department. Regardless of the countermeasure's effectiveness, resources were allocated.

Another credible measure of management commitment comes from setting objective measures in which safety can rate annual performances of the supervisor and department manager as well. Often, annual reviews are tied to salary increases, but regardless, having management-level associates self-rate in comparison with objective measures of safety performance is critical in establishing safety as a true virtue rather than a mere priority.

Management personnel have attitudes and perceptions about safety, the safety department, and its personnel. These attitudes and perceptions work to shape the picture of collaboration between safety and other organizational departments. Management attitudes and perceptions can be surveyed.

The competency level of management can also be measured from the amount of safety training and education completed. Programs can develop management-level safety training programs and can also use outside sources for professional safety development. Sources of safety-specific training can include Occupational Safety and Health Administration (OSHA) Training Institute Regional Centers, universities, professional organizations, conferences, consultants, and private companies. Programs can be specific in regard to sources and topics or use an overall points system for completion of training or attendance at professional conferences.

Compliance to company mission and guiding values, as well as to its own policies, and regulatory requirements also reflect the culture of the management in an organization. Individual as well as group decisions can be rated for compliance.

Employee safety culture has the same divisions for measure that have been covered for management-level personnel: participation, commitment, attitudes and perceptions, competency, and compliance.

Participation can be the allocation of workforce representation and attendance on safety related committees. But more important measures of representation can be tracked from the core management programs and the accuracy of employee reporting. A safety program should use a hazard-recognition program that has an employee-participation component. Conducting a job hazard analysis and soliciting employee input is not relying on employee participation. An example of reliance on workforce participation is a program where the workers on a specific line or process are empowered with the duty to recognize and report the level of hazards consistent with their competency. Higher-competency-level associates would then confirm results. It can be tracked in a such a way that, when safety conducts audits and general walk-through inspections, the discovery of a level 1 hazard, or a hazard that all associates can recognize and correct but that had not been reported would be tracked against the number of hazards reported in a timely manner and corrected. Results would indicate whether workforce associates were participating accurately in safety.

Accuracy of overall reporting is also an indication of commitment. The safety department must first document all reports and activities that come to its attention. This is done via an incident-control log, where each case or activity is assigned a unique identifier. Along with other data, the incident can be tracked as reported accurately or inaccurately. Accurate reporting would be factual and timely. Reports after a reasonable timeframe, inaccurate or false reporting, or a failure to report become tracked against accurate reports. A good example of failing to report an incident would be the finding of property damage that had no corresponding report. Because establishing the failure to report an incident is not easy and inaccurate reports are sometimes hard to establish, small measures of inaccurate versus total reports of incident indicate a commitment and participation issue. The organization will have to track data over time in order to measure improvement or lack of improvement in this area.

Employee commitment may be a combined measure of performance with core management programs such as hazard-recognition program reports of self-correction, behavior observations reports, employee safety awards, or performance from safety audits for level-one hazards (hazards all associates can recognize and correct).

The workforce's attitudes and perceptions toward their safety department and toward their own performances are an important reflection on whether safety is being established as a virtue. Typically, the workforce is surveyed. Surveys should protect anonymity and also be solicited in a manner that discourages hurried or token efforts. Passing out surveys at the end of meetings or near the beginning or end of a shift encourage inaccurate survey results. Surveys should be open, and other avenues of measuring attitudes can also be helpful.

Workforce competency is very important. All associates, regardless of workforce or management level, must have basic hazard-recognition and reporting training, otherwise associates cannot be empowered to participate in their own safety. But competency levels can progress, and outside sources can be used for professional development in safety. But positional levels such as maintenance mechanic level 1 and maintenance mechanic level 2 should have safety components when it comes to promotion as well as technical training. Workforce-level associates who work on safety committees require committee conduct training and education that introduces them to problem-solving methods. As their experience and education in regard to working in groups progresses, these become measures of competency as well.

Workforce-level compliance can be tracked using the number of disciplinary occurrences, such as verbal warnings, results from compliance walk-through inspections, causal analysis findings of unsafe acts, or from the findings of a mature behavior-based safety program.

The combined measures of management and workforce culture define the level of safety in the overall organizational culture. It can be tracked for comparison to category 3 metrics, or lagging performance indicators. The core programs that deliver services should be managed from the aspect of setting core metrics or metrics that will be reported on for measuring the overall performance of the safety program.

Metrics are divided into three categories. Category 1 metrics are those precursors to positive outcome that are leadership and management related duties or responsibilities. Category 2 metrics are cultural and operational precursors to positive outcome. Category 3 metrics are the bottom-line measures for performance (Mathis, 2014). Each core management program should have all three categories of measures that then provide a more complete picture of organizational safety performance as compared to only reporting on bottom-line statistics. Tracking initiatives in this manner will also allow the organization to tailor its efforts, concentrating on efforts that have the most positive impact on bottom-line numbers.

Each core management program will have category 1 metrics that should be identified for tracking. The safety manager will be responsible for setting which metrics will be tracked and reported on in order to demonstrate the program's level of success. Category 1 metrics can include the allocation of necessary tools, as well as other resources such as training. The duties supervisors, competent associates, or managers have in order to make the program effective provide a basis for determining level 1 metrics.

Each core management program will also have operational demands and produce attitudes and perceptions concerning safety. These identify the level 2 metrics. The application of procedure might be a level 2 metric for core programs and for day-to-day safety initiatives. For example, the display of correct procedure is very important for day-to-day lockout/tag out programs. The measure of proper procedures in a behavior-based safety program are also very important, just assessed differently. In a lock-out tag out (LOTO) program, safety professionals or competent level associates may observe the procedure or a number of procedures as a sample of what is occurring and provide a rating to the prescribed steps. In a Behavior Based Safety (BBS) program, the goal is identify why there is a gap between any designed procedure and actual occurrence so that they are as close as possible to being the same. Observations may be scored in a manner that provides a rating of congruency between the two for each machine. Both are examples of level 2 operational criteria. Other level 2 metrics center on perceptions and attitudes, either displayed or surveyed.

Category 3 metrics are the bottom-line measures for the program and for overall safety performance. Measures of injury, injury rates, or costs, for example, are just as important as leading metrics identified in categories 1 and 2. These bottom-line measures will be examined against the leading metrics in order to identify the program components with a positive impact on either program or overall performance. For example, suppose competent level associates are performing observations after peer-to-peer observations in a manner that confirms or provides balance to the scoring of the desired behaviors from the authorized level of observation: peer to peer. If we tracked the number of competent-level observations performed as a level 1 metric—because leadership is by example—we might compare the completion rate of competent-level observations to a rating of safety for the machine or to incidents to the machine. We would hope to find that the safety rating would be better, or fewer incidents occur on production machines where competent-level personnel are performing more observations. Maybe this would be due to leadership by example, more safety supervision, or more awareness by competent-level associates of the problems and corrections to those machines. But regardless, we could identify if the program components are having the desired impact.

Therefore, it might be helpful for the safety manager to identify level 1 through level 3 metrics for each core management program and for day-to-day safety-policy programs. Table 1.4 shows an example format for a tool that helps plan for management based on continuous improvement measures.

TABLE 1.4 *Example Planning Tool for Management*

Management Program	Category 1 Duties	Category 2 Operational Requirements	Desired Bottom-line Impact	Category 1 Metrics	Category 2 Metrics	Category 3 Metrics

In the above organizational tool, the safety manager can correspond metrics to program requirements and identify the tracking requirements that correspond to managing the program.

REFLECTION 1.3

1. Why is it important for policy to reflect organizational values?
2. Can you identify the most important safety virtue and justify why it is first?
3. Of the five categories for measuring safety, which is most important for measuring management culture? Employee culture?

COMMITTEE LEADERSHIP

The strategy for assigning responsibility for ethical oversight is akin to strategies recommended to avoid the groupthink phenomenon. Groupthink is an important concept to explore and practice preventive measures because it is detrimental to organizational decision making. Even when an organization's members share the same values and are mindful of organizational ethics, they will hold different opinions. When cohesiveness is high, pressures create a tendency to think or behave in a way that goes along with the behavior or thought pattern of the whole group. This results in a group that is willing to take more risk than would typically be taken by an individual. Groups tend to exclude certain members by process, assumptions, and group traits (Janis,1982). Collective decision-making failures are caused in many instances by members of the group who are reluctant to express unpopular views or to disagree with the popular view of the group. Irving Janis first used the term *groupthink* after examining how the "pressure to conform" influenced politics in the United States. Janis attributes groupthink to three causes: overestimating ability, closing of minds to information, and a desire to preserve group unity. A group's record of success tends to enhance this effect (Janis, 1982).

Symptoms of overconfidence in one's ability include discounting or ignoring possible failure scenarios, lack of contingency plans, and discounting or ignoring contrary data. Overestimation tends to encourage behavior described as a superiority complex, as well as decision making that is unethical in terms of personal and organizational values (Janis, 1982).

Being closed-minded consists of two symptoms: rationalizing based on past experience and negatively stereotyping outsiders. Those who rationalize risk based on past experience typically point out traditional methods as well as a lack of occurrence. Behavior that produces a "we versus the outsider" attitude as well as competitively discrediting others through criticism defines negative stereotyping (Janis, 1982).

The desire to preserve unity is strong. Self-censorship is common. Members devalue their own ideas and fail to bring up data with the group that could be of value. Direct pressure against any group member who presents data contradictory to group assumptions, stereotypes, and preferences discourages the exploration of options. It begins to encourage group members to self-guard against such information (Janis, 1982).

Groupthink leads to characteristic consequences. These seven consequences are 1) failing to identify an adequate number of alternatives, 2) failure to consider an adequate number of objectives, 3) failing to examine the risks of the popular choice, 4) failure to reassess initially rejected options, 5) limited researching of background information, 6) selective bias when assessing data and options, and 7) failing to have backup plans (Janis, 1982).

Janis describes the groupthink phenomenon as a "deterioration of mental efficiency, reality testing, and moral judgment" (1982, p. 9). The moral tie is of particular interest because it brings leadership into the solution. Leadership develops from moral virtue. Leadership to overcome groupthink in organizational decision making becomes central in safety management because participation is relied on for accomplishing the core duties of identifying hazards, assessing hazards, planning countermeasures, implementing countermeasures, and assessing the effectiveness of the counter, and making adjustments and standardizing successes. The core duties require group decisions.

In order to minimize groupthink, some general concepts guide behavior and policy. An open climate must be a value. Most organizations promote an open-door policy, but it is not backed up by training in presenting criticism and receiving criticism or negative information. People may adopt these unwritten courtesies due to group cohesiveness. But cohesiveness can create a sense of attack and ruin working relations between members. Diversity can help combat this, but only when members have set rules or guidelines on reporting and receiving information. Diversity in industry settings includes cross functional representation. Varying groups have diverse tasks and functions that can be impacted differently by a course of action.

Janis suggests the leader of a decision-making group should assign each member to be a critical evaluator and encourage seeking and presenting contradictory information, objections, and doubts. He points out that this requires more time for the decision-making process and that working relationships often suffer from the objections, doubt, and seeking of contradictory information (1982). This is why an open-door policy must be established and supported with training on reporting and receiving negative information and criticism. Committees must initially take on the task of establishing ground rules and duties for process and conduct.

Another of Janis's recommendations for an open climate is for upper-level managers to refrain from stating preferences and setting restrictive expectations when a group is formed for decision making. He also recommends setting up concurrent and independent groups to work on the same problem, with different members and leaders. In deploying this strategy, Janis recommends that guidelines address loyalty. Loyalty must be defined as applying to the organization as a whole rather than to one group (1982).

In order to avoid insulating the group from the rest of the organization Janis recommends three specific practices. When the group is assessing the feasibility and effectiveness of possible alternatives, Irving recommends that the group divide into two subgroups with different chairs. The subgroups can then reconvene and discuss their findings. The second practice is for individual group members to discuss the committee's deliberations and possible alternatives and repercussions. This is really similar to acting as a statesman. The final strategy is to bring in topical experts from within the organization who are not committee members to challenge the group's views (1982).

Overcoming leader bias begins with assigning a devil's advocate for each meeting. This person will present opposing data and question the popular choice. Janis points out that this role must be shared so that the same person is not continually assigned this role. Otherwise, opposition may be immediately discredited. If the decision-making group has an outside rival, considerable time must be spent identifying and assessing the signals and possible responses from the rival. It is important that the group does not harden itself against rivals by discrediting their potential. Finally, once the group has reached a seemingly final decision, a session must be scheduled for the group to come back and present a last-chance consideration of objections and alternatives (1982).

During the organization phase, the committee or group is formed. An organization may have policy guidance that dictates group membership by position. For example, if this was a committee looking to modify the workstation to align with human-factors issues, we might mandate that engineering, maintenance, safety, frontline supervision, workforce associates, and management join the committee. Recruiting or assigning committee duty is performed in different manners from volunteering, recruitment, or assigned participation.

Once initial forming takes place, the group decides on procedures and responsibilities and makes general assignments. One of those appointments is responsible for ethical oversight. This person will have the permanent assignment of briefing the others on organizational mission, values, and guiding principles and to report to the group on, or coordinate assessment of, the ethical adherence of alternative solutions the group considers. As alternatives are considered, or in discussion of ideas, the person responsible for ethical oversight must rate or assess each alternative for alignment with the organization's mission and values.

Procedurally, the group takes on the responsibility by each member to act as a critical evaluator to all the alternatives considered. The group leader or chair must make this a duty for all members since assigning a devil's advocate to each alternative will only be known between the leader and devil's advocate on a case by case assignment as alternatives are considered.

Perhaps most important at this stage is establishing what Irving termed "statesman duty" to all members. Talking about how data will be received and discussed in such a way as to discourage groupthink is important, as is how to speak with organizational members outside of the group that the member represents. Procedures can be explained. For example, one way to avoid dissent in the group is to require reports submitted to the group leader in advance of the meeting that detail criticism of the alternative. The leader can then summarize findings in a way that perhaps preserves some anonymity. Secret voting is another method of preserving anonymity. Ground rules play an important part in maintaining some group cohesiveness without encouraging groupthink too. Rules that require only factual data or a rule that forbids approaching members in accusatory manners or competitive approaches, such as pre-meeting campaigning for a personal preference, are examples that can prevent group dysfunction while not encouraging groupthink.

Once groups are formed, the project moves into the concept phase. Here the general purpose and requirements of the group are clearly defined. A clear definition of the problem must be expressed. For safety-related incidents, this usually means the investigation has been concluded through the causal analysis. If the group is forming to analyze for causation, the model used for analysis is established and applied. In safety incidents, the problem is the totality of the causal analysis. Statesman duty is important here because now group members can obtain input from members of the organization outside of the group, examine similar projects at the organization, identify outside experts, and generally prepare for a brainstorming session.

Once the group has identified optional alternatives, managing the group by assigning a devil's advocate, breaking into subgroups, assigning ethical oversight, and performance of critical evaluations become important in order to prevent groupthink. Once the concept stage has arrived at a preferred solution, the alternatives should be reexamined for benefits and risks, as well as for ethical alignment.

Groups now move the project forward to a prototype stage, testing phase, and production or implementation phase. Each cross-functional department must complete assigned duties before the project is passed from one stage and into the next. It is still important to assess the progression of the countermeasure, modification, or new machine or process for ethical alignment and performance. If either is unacceptable to the committee, then it must be corrected or discarded for another alternative as soon as possible.

REFLECTION 1.4

1. Can you think of practical examples or times that you have witnessed symptoms of groupthink?
2. What methods for countering groupthink can be applied to daily activities in meetings? Policy analysis?

CONCLUSION

Organizational management must establish leadership virtues that include safety at the moral level of its members. This governs individual behavior as well as group behavior. The culture must be measurable at the management and workforce levels. It defines the organization and places an identifiable potential to the organization's success. It is as though when you tug at one aspect of the organization, you find you are tugging at the moral fabric that defines what is valuable.

Indeed, the entire management must make it a goal to develop positive virtues within the entire organization so that its values are enforced with unwavering commitment. These values include safety. As much as possible, an organization's ethics must be reflected in the criteria of safety culture from the root level to the immediate level visible as the organization's symbolism. Transformational leadership serves as the driver that makes the workplace safety manager successful in forming relationships with members of the organization. Then it becomes a matter of allocating resources to efficiently address problems. This endeavor requires construction of a comprehensive management system.

REFERENCES

Aguayo, Rafael. *Dr. Deming: The American Who Taught the Japanese about Quality*. New York: Simon & Schuster, 1990.

Ciulla, Joanne B. *The Ethics of Leadership*. Belmont, CA: Wadsworth/Thomson Learning, 2003.

Gnambs, T., and B. Batinic. "The Roots of Interpersonal Influence: A Mediated Moderation Model for Knowledge and Traits as Predictors of Opinion Leadership." *Applied Psychology: An International Review, 62*(4) (2013): 597–618. doi:10.1111/j.1464-0597.2012.00497.x.

Hinsz, V. B., and G. S. Nickell. "Positive Reactions to Working in Groups in a Study of Group and Individual Goal Decision Making." *Group Dynamics, 8*(4) (2004): 253–264. doi:10.1037/1089-2699.8.4.253.

Howard, W. C. "Leadership: Four Styles." *Education, 126*(2) (2014): 384–391.

Ismail, A., N. A. Zainuddin, and Z. Ibrahim. "Linking Participative and Consultative Leadership Styles to Organizational Commitment as an Antecedent of Job Satisfaction." *UNITAR E-Journal, 6*(1) (2010): 11–26.

Janis, Irving, L. *Groupthink*, 2nd ed. Boston: Houghton Mifflin, 1982.

Kornør, H., & H. Nordvik. "Personality Traits in Leadership Behavior." *Scandinavian Journal of Psychology, 45*(1) (2004): 49–54. doi:10.1111/j.1467-9450.2004.00377.x.

Maslow, Abraham. (1943). "A Theory of Human Motivation." *Psychological Review*, vol. 50 (1943): 370–396.

Maxwell, John, C. *Developing the Leader Within You*. Nashville, TN: Thomas Nelson, 1993.

Maxwell, John, C. *The 21 Irrefutable Laws of Leadership: Follow Them and People Will Follow You*. Nashville, TN: Thomas Nelson, 1998.

Mathis, Terry L. "Common Practice: The Third Level of Leading Indicators." *EHS Today* (2014), ehstoday.com/safety-leadership/common-practice-third-level-leading-indicators.

McCrae, R. R., and P. T. Costa. *Personality in Adulthood*. New York: Guilford Press, 1990.

McGregor, Douglas. *The Human Side of Enterprise: Annotated Edition.* Annotated by Joel Cutcher-Gershenfeld. New York: McGraw-Hill, 2006.

Roughton, James, E. & James J. Mercurio. *Developing an Effective Safety Culture: A Leadership Approach.* Woburn, MA: Butterworth-Heinemann, 2002.

Ruggieri, S. & Abbate, C. S. (2013). *Leadership Style, Self-Sacrifice, and Team Identification.* Ruggieri, S. and C. S. Abbate. "Leadership Style, Self-Sacrifice, and Team Identification." *Social Behavior and Personality: An International Journal, 41*(7), (2013): 1171–1178. doi:10.2224/sbp.2013.41.7.1171.

Sandseter, Ellen and Leif Kinnear, Leif. (2011). "Children's Risky Play from an Evolutionary Perspective: The Anti-Phobic Effects of Thrilling Experiences." *Evolutionary Psychology Journal,* 9(2) (2011): 257–284, www.epjournl.net.

Schein, Edgar, H. *Organizational Culture and Leadership,* 4th ed. San Francisco: Jossey-Bass. 2010.

Wagner, Paul A. and Douglas J. Simpson. *Ethical Decision Making in School Administration.* Thousand Oaks, CA; Sage Publications. 2009.

Zagzebski, Linda T. (2004). *Divine Motivation Theory.* New York: Cambridge University Press, 2004.

Zehir, C., Müceldili, B., Altindağ, E., Şehitoğlu, Y., and S. Zehir. Charismatic Leadership and Organizational Citizenship Behavior: The Mediating Role of Ethical Climate. *Social Behavior and Personality: An International Journal, 42*(8) (2014): 1365–1375. doi:10.2224/sbp.2014.42.8.1365.

Figure Credits

Fig. 1.1: Adapted from Douglas Simpson and Paul A. Wagner, Ethical Decision Making in School Administration. Copyright © 2009 by SAGE Publications.

CHAPTER TWO

How to Structure the Safety Management System

FOREWORD

Successful safety management requires leadership to set culture and the efficient allocation of resources in response to challenges to workplace safety. Safety management is the act of controlling the process of identifying safety challenges, investigating them for causation, formulating and implementing countermeasures to the challenges, and then adjusting for efficiency. At the heart of managing is measuring. Managers must measure the process of safety. Leadership influences, then management efficiently directs resources. Both leadership and management are vital to a successful safety program in any organization. Figure 2.1 depicts the relationship between leadership and management in regard to success.

A safety management system ensures the sustainability of the organization's learned safety lessons. It maintains a level of reassurance to those it serves that they can predict an ethical and just outcome with consistency. It helps to form the bond of trust so important in the process of influence. A system also serves the management effort, or the control of processes. Safety management must allocate resources efficiently in order to maintain credible trust with members at all levels of the organization in order to be considered successful. Generally, safety must allocate its resources where they are needed most in order to fulfill its obligations.

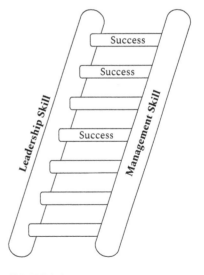

FIGURE 2.1 Ladder

Objectives

After reading this chapter the learner will be able to:

- Author a philosophy of safety management
- Assess needs of the safety management system based upon management philosophy
- Assemble safety-program elements based upon organizational need
- Assemble a safety operations plan
- Assemble program-effectiveness measures

LEARNING PLAN

LEVEL	ULTIMATE OUTCOME	CLAIM	LEARNING TYPE	ASSESSMENT
U	Author a philosophy of overall safety management	Managers of workplace safety must have an overall philosophy that guides efforts to control the processes involved in safety management.	CA/CT	Students will develop a visual model of workplace safety management to reflect core duties, customers of safety, and services provided to customers. Students will develop a management philosophy that will serve as the ethical check for controlling processes involved in safety management.
U	Assess needs of the safety management system based upon management philosophy Assemble safety program elements based upon organizational need	Assessment of the scope of the safety management system begins with matching customer needs to delivered services.	CA/CT	Students will identify core management programs that deliver services to the customer.
U	Assemble a safety operations plan.	Written safety operations plans are critical for sustainable management efforts.	CA/CT	Students will identify and formulate the management section of a safety operations plan.

	Assemble program effectiveness criteria	The practice of continual improvement requires core management programs and compliance programs to be assessed for effectiveness.		Students will design a program effectiveness audit based upon metrics structure for a core management program or safety compliance program.
U			CT	

Learning Plan Legend

Level:

F: <u>Foundational Outcomes</u>: basic abilities

M: <u>Mediating Outcomes</u>: Progress through a developmental model; interpret, analyze, evaluate progressively challenging claims, arguments

U: <u>Ultimate Outcome</u>: Navigate most advanced arguments/claims

Type of Learning:

CR: <u>Critical Reading</u>: The ability to read, process, and understand the meaning of written information

IL: <u>Information Literacy</u>: Locating and selecting suitable information for a task; evaluating appropriateness/validity of information sources

AC: <u>Application of Concepts</u>: Ability to apply discipline-specific knowledge/skill to tasks/situations important to the discipline

CT: <u>Critical Thinking</u>: Ability to apply a concept to a vague or argumentative claim without a creative leap.

AT: <u>Analytical Thinking</u>: Ability to critique/analyze situations using a concept or model.

CA: <u>Creative Application</u>: Ability to apply a model/concept in a new way/to an unrelated situation or scenario. Involves creative leaps.

DETERMINING THE SCOPE OF THE SYSTEM

Just as performing a hazard inventory is important for understanding the scope of hazards and regulations that must be dealt with, identifying the overall scope of the safety management system is equally important. As the hazard inventory reveals the extent of compliance programs that must be administered, the identification of the services that safety must provide to its internal and external customers is revealed from analysis of the customer's needs. In order to form a view of the scope of the safety management system the manager must have an idea of an overall management philosophy. In the 1960s, public safety began adopting Community Oriented Policing as a methodology for managing public safety. It is a philosophy that mirrors many management efforts with workplace safety and can be adopted for private sector use.

A typical image of a regulatory agency is that the agency itself is outside of the circle of the industry that it is meant to regulate. It becomes strictly a slapping-of-the-hand entity that overlooks the industrial community. The safety department of an organization simply cannot operate in this manner. Safety must be a problem-solving partner rather than oversight.

Community Oriented Policing viewed police agencies as being central and equal to the community circle. Other subdivisions of the community, such as the business sector, schools, civic organizations, and neighborhoods had problems and needs that the police agency could address. Since the community subdivisions were equal, the police agency would form partnerships with them in order to identify the problems, solutions, and services that the agency could provide to them. Common examples like Neighborhood Watch, D.A.R.E., theft prevention courses, school safety programs, and citizen academies emerged and served to build trust through transparency and participation.

Just as COP serves the cross functional division of a community, Employee Oriented Safety, the mirror of COP in a private sector industry, serves the cross functional division of an industrial organization. Safety relies on partnerships with the workforce associates, labor organizations, management, engineering and maintenance, and with external customers like vendors, ground guests, contractors, neighbors, first responders, and employee families, just to name a few possible customers.

COP applied as Employee Oriented Safety in a private organization places a priority on injury reduction but makes clear that safety serves internal and external customers of the company. This means that safety as a cross-functional division of an organization is an open system in that it must rely on input and feedback from components outside of safety itself. The core duties of safety are to: 1) identify negative and positive trends, 2) investigate the trends to plan countermeasures to problems, 3) implement the countermeasures, and 4) assess the effectiveness of countermeasures and adjust them for improvement. The visual picture of the safety department is shown in Figure 2.2 to follow.

From the visual model that identifies customers we can begin to identify services that safety is capable of delivering and as the program develops, it can identify further needs of the customers, and, if resources are available, fill those needs. It also provides a view of what partnerships can or need to be developed. This is the beginning of building a system that meets the scope of the needs of the organization. Figure 2.2 demonstrates the open environment, or interaction with external components, of safety and the customers that it serves.

This system is the way that the organization learns workplace safety. Therefore, core duties are extended to communicating the learned information. In fulfillment of its primary concern of injury reduction, it manages loss.

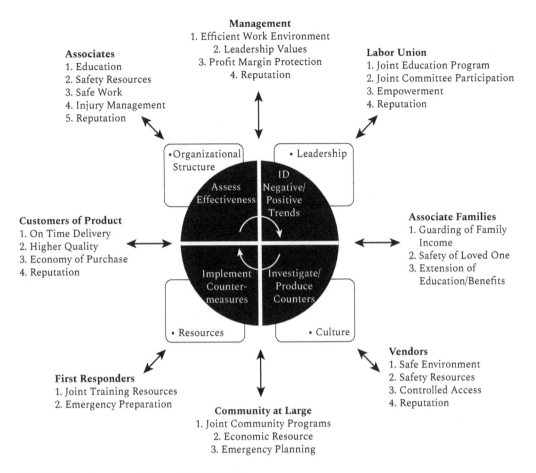

FIGURE 2.2 Employee Oriented Safety Model

Safety management begins by developing core management programs that deliver services to customers by accomplishing core duties. Core management programs allow the system to deliver services to customers. Core management programs are different from safety initiatives, or the day-to-day compliance programs that the hazard inventory identifies the need for. These day-to-day policy and procedure programs or compliance programs become supporting programs.

Table 2.1 provides an organizational structure for the formal components of the safety-management system based on core duties and customer needs. By analyzing customer needs, safety services can be identified and cross-matched to core programs, which in turn also have supporting needs. This is the method for determining if the safety management system scope is complete.

TABLE 2.1 *Safety Management System Organizational Structure*

CORE DUTIES	CORE MANAGEMENT PROGRAMS	SYSTEM SUPPORT
Gather Facts	Reporting Procedures	Incident documentation/logs/records
ID of Hazards, Threats, Vulnerabilities	Hazard Recognition Programs: 1. JHA 2. Employee Recognition Programs Behavior Based Observations/Compliance observations Auditing Programs Early Intervention in System Safety Analysis Teams Emergency Response planning Security Assessments	Program Effectiveness Audits Metric Tracking Formal policy/procedures Training
Assessment of Hazards	Investigation programs System Safety Committees Security Reviews	Program Effectiveness Audits Metric Tracking Formal policy/procedures Training
Planning Countermeasures	System Safety Analysis Committees/Problem Solving methods Emergency response/First Aid Programs	Program Effectiveness Audits Metric Tracking Formal policy/procedures Training
Implementing Change	Organizational Change Plans	Program Effectiveness Audits Metric Tracking Formal policy/procedures Training
Reassessment		Program Effectiveness Audits Metric Tracking Formal policy/procedures Training
Organizational Communication	Production of company standards	Program Effectiveness Audits Metric Tracking Formal policy/procedures Training
Injury Management	Return to Work Programs Safety Away from Work programs (SAWs)	Program Effectiveness Audits Metric Tracking Formal policy/procedures Training

Employee Oriented Safety begins with a philosophy that considers the customers it serves to be equal in standing; that customer input and participation are required for system function; that partnerships with customers are necessary for problem identification and solutions; and finally, that the scope of the system is determined from customer needs . Employee-oriented safety aims to develop competency to the point that actual empowerment takes place. Duties are delegated down as far as possible.

The management philosophy is as important to management as guiding values are to leadership. If the philosophy is communicated and exemplified to members of the organization, it ensures sustainable management practice in the same way as organizational ethics ensures trust in safety leadership. Members can predict their roles, management procedure, and outcome consistently. Transparency through active participation builds trust.

Figure 2.3 consists of a formal communication to the membership of overall management philosophy based upon the application of Community Oriented Policing to organizational safety. In review of the concept, the cross-functional department of safety provides services to varying subdivisions of the organization and to peripheral subdivisions of the community in which the organization resides. These services provide the skeletal base for determining the core management programs that deliver services to the various customers. Each customer is relied on to actively participate in the core duties of safety to form partnerships with safety that identify problems and participate in formulating solutions and implementing policy. Transparency and actual empowerment build a level of trust that is foundational for the system to function.

Production *without* **Safety** and **Quality** is not production, it is merely *activity*.

Production + Safety + Quality = Success

Safety is not a priority. It is a virtue that cannot be waived.

TABLE 2.2 *Safety Virtues*

COURAGE	To do the task safely
	To report hazards
	To correct hazards
	To correct undesired behavior
PATIENCE	Not to take a shortcut
	To consult with task experts
	To consult with others
HONOR	To hold yourself and others to work safely even when no one is watching
LOYALTY	To the safety of your fellow associate

RESPECT	For the well-being of self and others
DISCIPLINE	To not prioritize safety
HONESTY	In safety participation
COMPASSION	For new hires and returning casualties

Mission: Safety will empower the members of XYZ Corporation to practice production with a virtue of safety by providing the necessary resources for safety partnership.

Empowerment: TASK = transparency acknowledgment and safety knowledge

Here at XYZ Corporation, safety is a partnership that relies on all of its members to actively participate to identify problems and find effective solutions.

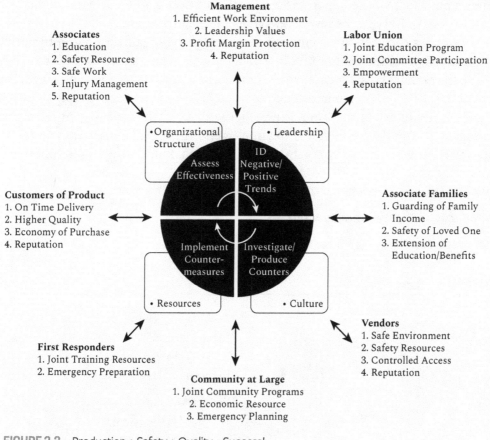

Management
1. Efficient Work Environment
2. Leadership Values
3. Profit Margin Protection
4. Reputation

Associates
1. Education
2. Safety Resources
3. Safe Work
4. Injury Management
5. Reputation

Labor Union
1. Joint Education Program
2. Joint Committee Participation
3. Empowerment
4. Reputation

Customers of Product
1. On Time Delivery
2. Higher Quality
3. Economy of Purchase
4. Reputation

Associate Families
1. Guarding of Family Income
2. Safety of Loved One
3. Extension of Education/Benefits

First Responders
1. Joint Training Resources
2. Emergency Preparation

Community at Large
1. Joint Community Programs
2. Economic Resource
3. Emergency Planning

Vendors
1. Safe Environment
2. Safety Resources
3. Controlled Access
4. Reputation

• Organizational Structure • Leadership • Culture • Resources

Assess Effectiveness — ID Negative/Positive Trends — Implement Countermeasures — Investigate/Produce Counters

FIGURE 2.3 Production + Safety + Quality = Success!

Participation occurs in the safety system in different ways. Participation can occur in active safety committees that are standing and perform a variety of duties. Examples include committees that assess processes for hazards, assess production lines and activities for human factors or ergonomic improvements, system safety committees that look at new machinery or modification to existing machines, or regular committees that serve to communicate and train the body of the workforce. Participation can also occur in the form of ad hoc or as-needed committees when special issues arise. Participation can also occur as delegated duties of performance based upon competency. But in the Employee Oriented System, participation is not merely ordering associates to act. Participation has the elements of developed competency, authority to act at some level, and oversight that assesses the performance level of participation. This becomes clear when we examine a three-tiered hazard recognition program in chapter 4. One tier of the program relies on workforce competency to perform a core duty of the safety function and the performance is measured for accuracy and helps determine culture. This is an important aspect to equality of service.

Empowerment means that associates may act or make safety decisions in accordance to the level of their competency. Employee Oriented Safety (EOS) must develop competency in order to facilitate participation and delegate authority to act and make decisions. The safety program and all of its core management programs and initiatives or compliance programs have three levels of competency that are tied to responsibilities. The first level is the "authorized" level. Here base competency is developed and base responsibilities are assigned. Common to all programs and initiatives are the responsibilities to recognize and report hazards, vulnerabilities, and threats associated with the program and to follow proper safety protocols. The "competent" level is next. This level assumes the competency needs and responsibilities of the first level and is challenged with more advanced duties. Common to the "competent" level is the competency of hazard abatement, basic investigational skill, and authority to enforce. Responsibilities would then be to assist in or plan abatement, enforce policy, and assist with investigations. The administrator level is responsible for the entire endeavor and must measure the program and implement improvements.

The organization of metrics is another key element of overall design and scope of the system. The general responsibility of any manager is the overall success of the process(s) he or she is responsible for. This means managers allocate resources at the point of need that will produce the most positive effect. Simply said, this means the efficient allocation of resources—managers must measure. Every core management program and even supporting compliance programs must have measures so that negative and positive trends can be identified and improvement can take place continuously. EOS uses a three-tiered structure of management metrics.

Level 1 metrics are those measures of leadership and management responsibilities. The premiere example is the ratio of closed audits or countermeasure plans to open or incomplete ones. This measure indicates the level of management commitment because of the allocation of resources. Level 2 metrics are measures of culture or perception, beliefs, and priorities. Level 3 metrics are performance metrics typically referred to as lagging indicators (Mathis, 2014). Each core management program and compliance program is assessed by comparing level 1 and 2 metrics to performance indicators.

REFLECTION 2.1

1. Why is the efficiency of allocating resources a determining factor of success for a safety manager?
2. How does the extension of safety to the community benefit the organization overall?
3. Is transparency the link between management and leadership? Why?

TRACKING SAFETY ACTIVITY

The management system must be comprehensive. This means that it:

1. Collects relevant statistics and information (facts)
2. Organizes and analyzes data (investigates)
3. Counters or abates problems and produce standards
4. Monitors changes
5. Communicates with all the components or customers

Guiding the functions of the system are two foundational principles. The first is early intervention. Early intervention is the concept that the sooner a problem or negative trend can be identified and corrected, the less costly it will be in terms of resources. Herbert Heinrich first touched on this concept in the early 1930s with his famous ratio of near misses to major injuries. He postulated that for every major injury, which we interpret today to mean an injury meeting OSHA recordkeeping criteria, there were twenty-nine minor injuries occurring, and three hundred near misses (Heinrich, Peterson, & Roos, 1980). More recent research has been unable to debunk this ratio, and some have found near misses to be much higher in ratio (Manuele, 2002). The second concept is that of "zero defects" as suggested by Philip Crosby. In using this concept we will not strictly advertise zero injuries as a goal due to motivational concerns over when an injury or incident does occur, and the negative impact it has on associates' reporting of incidents and injuries. Instead we will use this concept to make our system of gathering and analyzing information a system that does not exclude information that would identify negative trends. For example, we will define and assess near misses for causation and counters, or fixes as well as incidents that produce damage or injury. This begins with defining the activity the safety department will manage.

An occurrence is a factual happening. Occurrences can be those that have negative potential or actually result in negative consequences. An occurrence can be neutral or positive. Positive occurrences as far as safety can be those actions or activities from all cross-functional or organizational activities that result in production, or activity that produces product with quality and safety; production = product + quality + safety. Negative occurrences or incidents result in negative or undesired consequences, such as injury. For now we will track incidents and activity from the safety department. Later we will discuss positive occurrences.

The category of safety department activities can vary, depending on the structure of the organization. Activity categories can be organized as safety activity, environmental safety, or security, for example. Common to all the categories of responsibility that the safety department might be tasked with is whether the activity is proactive or reactive. Types of activity usually include incident investigations, audits, observations, committee meetings, regulatory inspection, meetings, strategic plans, sampling, or a host of other activities, depending on how the organization names and organizes specific tasks. Naturally, this depends on the customers' needs and programs.

Incidents will vary and will require further description because the nature of the incident is important in tracking trends. Incidents are factual events that have occurred and become known to the safety department. They are distinguished from activity in that activity is ordered or originates from the safety department. Clearly, once an incident occurs, safety directs activity, but the activity has originated in response to an incident. This is an important distinction in documentation because each incident or activity originating from the safety department will be issued a unique identifying number. For example, if a worker fell from a ladder and received first aid, this incident would be assigned a unique identifier. The investigation activity that corresponds to the incident will not receive a different identifier, but use the same identifier to connect all activity and forms associated with the event. Conversely, if the safety manager ordered all ladders to be inspected, but not in response to a tracked event, this activity would be issued a unique identifier so that all forms and subsequent activity can be connected to the original action. Thus the incident control log becomes necessary.

The incident control log gathers initial data about an incident and serves as a master log. This is important to safety managers because it shows the amount of activity handled by the department and provides justification for additional safety resources, such as new personnel. The log is also used to identify negative trends and prioritize resources based on the information collected. Information is gathered and coded in the log in such a way as to minimize its size and serve as a quick reference guide. We can use numbers to code incident control information, as well as colors to denote the activity category. Therefore, a legend proves useful. For example, we could define an incident control number as beginning with a certain letter that identifies the activity category. We may also use a color code for each activity category. For example, if the safety manager wanted to address security and environmental health as well as occupational safety, the incident control log could code all security-related activity as blue, safety-related activity as yellow, and all environment-related activity as green. The incident control number (IC number) could further describe the activity type. For example, suppose we add a letter in front of the unique identifier to denote the type. Safety activity includes incident investigations, audits, and committee meetings, for example. So, we could assign an A in front of any IC number to denote an audit. This way, by using color coding and unique identifiers, we can see the category of activity and the specific activity quickly. The concept can be applied to any organization's needs to collect whatever information the safety manager needs. Figure 2.4 reflects one possible example of an incident control log.

IC Number ????	Date/Day	Type Incident	Involved Perso	Superviso	Injury/: (Y/N)	Investigatio (Y or N)	Dept	Outside Agenc	Open/Close	Notes:
??0001	1/1/Tue	Alarm/Entry	N/A	N/A	N	N	Office	PD	Closed/Fals	Pres. Schumann entered
??0002	1/1/Tue	Alarm/Power	N/A	N/A	N	N	All	AEP	Closed/Fals	Power outage
??003	1/2/Wed	Injury	Randall, Dave	Dixon	Y/Y	Y	Weld		Closed	
??004	1/3/Thur	Prop Loss	Fee, Steven	Sears	N	Y	Ship	Eagle Trucking	Open	Forklift into trailer
??005	1/5/Sat	Near Miss	Chadwell, Shar	Sears	N	Y	Ship	N/A	Open	Forklift rolled out bay door
??006	1/6/Sun	Injury/Burn	Thomas,Scott	Simms	Y/Y	Y	Weld	N/A	Closed	
??007	1/7/Mon	Near Miss	Welker,Lisa	Ford	N	N	Ship	Eagle Trucking	Closed	Forklift almost hit by truck
??008	1/7/Mon	Near Miss	Moore, Ron	Dixon	N	Y	Weld	N/A	Closed	machine malfunction
??009	1/7/Mon	Theft	N/A	Gene	N	Y	Maint	N/A	Open	tool theft
??010	1/8/Tues	Audit	N/A	N/A	N	N	Weld		Open	machine guarding

FIGURE 2.4   Incident Control Log
*Names are fictional and any similarity is unintended.

From the above figure of the incident log, let's examine the IC number a little closer. The first two letters can give the specific type of activity, and the first two numbers indicate the year of occurrence. The next set of numbers after the dash indicate the number of incidents in order of occurrence. We can place a code at the end that indicates actions taken. For example, it would be convenient to know if an injury incident had a worker's compensation claim associated with it. The letters WC could indicate this action. Other codes could be attached to indicate other actions that need to be tracked. An example of a complete IC number might appear as: SAF17-003WC. Thus, this IC number tells us that the third incident of 2017 was a safety-related activity with a worker's compensation claim filed. The other data blocks can indicate much more information, as revealed in the above example. However, the incident log is not the final stopping point for data.

Once the incident is logged, the appropriate activity occurs. Depending on the activity and the actions taken, data may need to be entered into a separate log of statistical data. For example, suppose an incident involved an employee injury. Suppose this injury met the recording criteria from 29 CFR 1904, which are:

1. Work relatedness
2. New case or significant re-aggravation
3. Meeting one or more of the following criteria
   a. Death
   b. Day away from work
   c. Restricted work or job transfer
   d. Loss of consciousness
   e. Significant injury or illness diagnosed by a physician or other licensed health care professional
   f. Medical treatment beyond first aid

The case would have to be logged on the company's 300 log. However, we will need a separate injury log to track data specifically important to the company. The machine, process, line, and he part being produced would be significant information to track. These statistics alone allow us to determine costs, assess risk, prioritize assets, and include safety costs in future bids—all from tracking the history or experience per workstation. Figure 2.5 is an example of an injury log that tracks organizational experience surrounding injury incidents.

Date/Day	Type/body	Employee	Department	Shift/Supervisor	IC#/Open=Closed	Machine/Process	Part#	Immediate Cause	Claim#	Cost
1/2/????	Cut/finger	Randall Dave	Weld	2/Dixon	??003/Closed	AS 134	1z7682	gloves/sharp edge	DS??0001	880
1/6????	burn/finge	Thomas,Scott	weld	1/Simms	??005/closed	AS116	1h1324	hot part/gloves	DS??0002	668
1/9????	cut/hand	Sara,Connor	stamping	3/Donta	??012/closed	Press 10	1z7682-1	sharp edge/foot haz	DS??0003	886
1/15/????	Pinch/fing	Legg,Dan	shipping	2/Cheung	??014/open	Package	1cnhtong	heavy/lift procedure		60
1/18/????	cut/wrist	Cope,Brenda	coating	1/Alexander	??016/closed	offload	1z7682F	sharpedge/ppe	DS??0004	960
1/20/????	cut/finger	Dans, Rod	stamping	3/Donta	??019/closed	Press 10	1z7682-1	sharp/gloves	DS??0005	880
1/23/????	cut finger	Nantz, Rick	weld	1/Simms	??021/closed	AS 144	1h67S2	sharp/gloves		17
1/25????	sprain/wri	Hopkins,Tim	stamping	2/Blevins	??023/closed	Press6	N/a	surface	DS??0006 open	
1/27????	splinter/ha	Simms, Brad	Weld	2/Dixon	??027/closed	AS 117	1n76423-1	burr/ppe		7
1/28????	cut/hand	Wells, Lisa	Weld	1/Simms	??029/closed	AS210	6z9876-2	sharp/gloves	DS??0007	880
1/31/????	cut/finger	Barber, Sheri	stamping	1/Lucas	??031	Press 05	7zx7653-2	burrr/gloves	DS??0008	880

Yellow: Recordable
Blue: First aid only

FIGURE 2.5 Injury Log

Typically, logs are created around key topics for the management of safety, depending on the organization's activity and needs. Core management programs require logs in order to measure trends for continuous improvement. Any initiative that will be reported for measuring program success, or any topic that needs to be controlled to meet the safety department's mission must be tracked with a log. Examples of other logs are the behavior-based safety program log, the hazard recognition program log, the causal analysis log, and logs associated with measures for organizational culture.

REFLECTION 2.2

1. What are the positives and negatives of establishing "zero" injuries as a goal?
2. What does the management operational standard of "zero" mean to the design of the safety management system?
3. Why is measuring so important to managers?

ELEMENTS OF FORMAL MANAGEMENT SYSTEM STANDARDS

Safety management standards exist to give common structure to the management system for a wide array of applications. From a manager's point of view, the organization's safety management system provides a clear path to the safety activities required to fulfill customer needs. It allows the manager to focus resources in the most efficient places. From a leadership point of view, the system's managed programs must reduce the amount of risk members of the organization are willing to accept (Roughton & Mercurio, 2002). Having a clear picture of the common elements mentioned by safety management systems from OSHAS 18001, to Z-10, as well as the forthcoming International Safety Management System Standard, ISO 45001, provides guidance on the development of system processes.

Management systems for safety help an organization develop relevant policy; set safety goals; develop specific processes that identify and counter hazards, threats, and vulnerabilities; and evaluate performance of safety's core duties, while taking into account its legal and moral obligations to the members of the organization. Its success progresses with maturity. The development stage is where an organization develops the core management programs, supporting programs, and system supporting processes. Here the measure of success is the completion of this development and its implementation. The compliance stage is where the system is operating and measures of success depend on compliance to regulations, as well as organizational policy and procedures. The results stage is measured for performance. In this stage, level 3 performance indicators become important. The hazard elimination stage is when performance goals are being realized and the organization has matured to the point that one can measure the elimination of hazards as a measure of success.

In this stage, level 1 and 2, leading indicators become important specific measures (Kausek, 2007).

Overall indicators of success for the safety management system center on low occurrence level 3 performance incidents, low occurrence of new hazards, high levels of organizational compliance, high levels of organization attitudes toward safety, low levels of expenditures for legal liability, low levels of worker compensation expenditures, and other operational and financial benefits of proper safety management.

The first common element is that the system's scope must meet the needs of the organization. Based upon customers' needs and the overall management philosophy, the core management programs and their complexity become apparent. In turn, the supporting processes and the support requirements for the system as a whole are determined. Core management programs allow services to be provided to customers through completion of the core duties required by the customers. Table 2.3 as copied on the following page from earlier in this chapter provides insight to one possible skeleton of a safety management system (SMS) for one possible situational need.

TABLE 2.3 *Safety Management System Organizational Structure*

CORE DUTIES	CORE MANAGEMENT PROGRAMS	SYSTEM SUPPORT
Gather Facts	Reporting Procedures	Incident documentation/logs/records
ID of Hazards, Threats, Vulnerabilities	Hazard Recognition Programs: 1. JHA 2. Employee Recognition Programs Behavior Based Observations/Compliance observations Auditing Programs Early Intervention in System Safety Analysis Teams Emergency Response planning Security Assessments	Program Effectiveness Audits Metric Tracking Formal policy/procedures Training
Assessment of Hazards	Investigation programs System Safety Committees Security Reviews	Program Effectiveness Audits Metric Tracking Formal policy/procedures Training
Planning Countermeasures	System Safety Analysis Committees/Problem Solving methods Emergency response/First Aid Programs	Program Effectiveness Audits Metric Tracking Formal policy/procedures Training
Implementing Change	Organizational Change Plans	Program Effectiveness Audits Metric Tracking Formal policy/procedures Training
Re-assessment		Program Effectiveness Audits Metric Tracking Formal policy/procedures Training
Organizational Communication	Production of company standards	Program Effectiveness Audits Metric Tracking Formal policy/procedures Training
Injury Management	Return to Work Programs Safety Away from Work programs (SAW's)	Program Effectiveness Audits Metric Tracking Formal policy/procedures Training

This arrangement reflects the covering of the needs of many safety management departments across a variety of industry settings. Examining customers and their needs, both internal and peripheral to the organization, determine the overall scope of the safety program. Covering customer needs is the way an organization certifies that its safety management system meets the scope of the need. In the real world, resources are always limited. Organizations examining their scope, implementing a new system, or modifying their system should prioritize implementation of programs that will fulfill customer needs. The first priority must be the reduction of internal loss.

Management has certain duties under management-system standards. Management commitment and review is another common element in SMS standards. Management is responsible for the success of everything to do with the organization. Management, and in particular the general manager, president, chief of operations, or whatever the title may be, is responsible for setting the organization's mission and values, as well as determining overall strategic direction, determining the guiding strategy, assessing overall performance, and implementing change (Crossan, Rouse, Fry, & Killing, 2009). Specific to the safety management system, the organization's management collectively must ensure that the organization's activity, knowledge, history, and all risks are considered. Next, management must evaluate the strategic plan for safety and ensure that it is in line with the overall organizational strategy. Third, management must ensure that the core duties of safety are being completed with systematic support. All business processes must have safety incorporated into their policy and procedures. Further responsibilities include evaluating safety performance; allocating resources; ensuring team inclusion; and participation in safety activity and including safety into all business processes, such as business continuation planning. Promoting safety as integral to the organization's culture, as well as promoting continuous improvement are also direct responsibilities. Finally, management is responsible for developing safety knowledge and perceptions in all employees.

Hazard identification is the third common element of system standards. Identification process must include those applying to internal employees, any ground guests such as vendors and contractors, to those outside of the organization in the community who are impacted by the activities of the organization or possible emergency situations. This process must consider frequently encountered situations as well as less typical situations in order to be comprehensive. It should also include a proactive effort to examine recent studies, occurrences, and developments in the awareness of new hazards. This process should include an assessment of the legal liability of the organization in regard to identified hazards, both regulatory legal concerns and civil concerns. Rating or prioritizing risks is also an important piece of allocating resources.

Acceptable risk is a concept that drives prioritization. Acceptable risk really means that the addition of another control does not reduce its overall risk rating (ANSI B11.1, 2008). After a hazard inventory is completed, the hazard recognition program can assess risk based upon probability, frequency, and exposure with controls in place. Importantly, continuous improvement and acceptable risk are concepts that cannot be used to justify incremental progression toward basic compliance. Generally, abatement strategies resemble the common model of controls that utilize elimination and substitution, administratively controlling and limiting exposures, and providing personal protection.

Abatement of hazards, threats, and vulnerabilities must also be strategized and placed into written policy. System safety analysis policy is a formal effort to ensure total team representation for identifying production, quality, and safety problems early in the process of planning new launches of machines or processes or the modification of existing machines and processes, in order to include any workstation procedures. System safety analysis committees are also considered as process safety committees (formal processes by definition) and so are human factor or ergonomic committees that routinely, retroactively, or proactively assess the organization's activities, production lines, or processes.

It goes without saying that two-way communication methods must be used to identify and communicate hazards with internal and external customers of safety to increase awareness and competency of the organization's safety management efforts and of hazards and controls. Employee representation is typically a fourth common element to safety management systems. All levels of the organization should be included in committee assessment and planning efforts. This can be extended to the peripheral customers of safety present in the surrounding community to the extent that voluntary efforts exist. System safety analysis endeavors and communications combine to help organizations manage change.

Emergency response plans and investigational programs that include proactive audits are additional programs mandated by safety management system standards and must include the elements covered previously. Management commitment, associate participation, consideration of common hazards and incidents along with nonroutine or "what-if" emergencies, management review, and measures for continuous improvement efforts are necessary elements for emergency planning.

Performance evaluation for continuous improvement must include measures for meeting organizational policy and objectives, as well as measures for organizational ethics or adherence to organizational values. Controls must be measured and evaluated for improvement, proactive measures or leading indicators, and precursors of positive safety performance must be included in evaluation of performance.

REFLECTION 2.3

1. What is the most beneficial aspect of meeting a safety management system standard?
2. How does a safety management system standard help managers allocate resources effectively and reduce the amount of risk organizational members are willing to accept? What gaps does leadership fill?

STRUCTURING THE SAFETY OPERATIONS PLAN

A safety operations plan establishes a written record of how safety will be managed. As the general manager of safety your duties are no different from the duties of any general manager, just correlated to safety with a role to play within the established direction and strategy set by upper-level management. A general manager is responsible for setting direction, determining strategy, implementing changes, and assessing overall performance (Crossan et al, 2009). The plan's structure can vary, but it should incorporate the elements of a safety management system standard. At a minimum, the sections are:

Leadership Section
Vision statement/component
Mission statement/component
Guiding values
Safety philosophy
Safety culture

Management Section
Goals and objectives
Core metrics
Strategic plans

Core Management Programs
Communications
Hazard recognition
System safety analysis
Human factors
Investigations
Auditing
Injury management
Procurement

The vision statement for safety covers a performance level and validation measure for future accomplishment. It must match the organization's strategic plan and must be supported by upper-level management. It provides a vision of excellence and motivation without being unobtainable. It identifies the organization, but it does not tell the "what" or "how" of the mission. A simple example for safety is that the organization will reach and maintain a level of safety virtue, setting a standard for all in the community and industry. Establishing the true meaning of the vague statement is the responsibility of the safety manager.

The mission statement is more direct in application than a vision statement. It will mention the organization, its primary activity, who it serves, and the strategy used to complete the mission. The mission statement is where strategic planning and marketing merge, making it an expression of the organization. In regard to safety, we must go back to our visual model of what safety is and who it serves. We might formulate a mission statement that goes something along these lines: "The safety department here at XYZ Corporation will strive to facilitate a safe and healthful work environment for its associates and the surrounding community by providing the necessary resources to empower those it serves to make safe decisions." After the statement, we include further explanation making clear that resources include knowledge to recognize hazards; partnership to identify solutions; as well as hard resources, such as equipment.

The mission statement gives the organization's philosophy of how safety will be managed. It is the basis for ethical assessment. Beginning with the mission statement, safety will have established, guiding values. These should include leadership traits as well as nonleadership traits that describe how decisions should be made. Examples of guiding values include integrity, commitment, courage, compliance, corporate responsibility, and equal access. It is up to the safety manager to describe these terms and give examples in an effort to assign meaning to the guiding values and to ensure that policy and decisions are carried out consistently with values.

The guiding values become symbols of the organization and permeate its culture. These guiding values also become one of the two pillars of empowerment. Empowerment cannot exist without basic knowledge and guiding values. Guiding values are a guarantee to associates at all levels that decisions are made consistent with values; are fair decisions; and regardless of outcome, the decision will be upheld without consequences. Consistent with the first principle of leadership—leadership is by example—organizational ethics must be exemplified by management.

Safety philosophy in the leadership section defines the concept and even reveals how safety will be managed. This has been covered in this chapter.

Production + Safety + Quality = Success!
Safety is not a priority. It's a virtue that cannot be waived!

TABLE 2.4 *Safety Virtues*

COURAGE	To do the task safely
	To report hazards
	To correct hazards
	To correct undesired behavior
PATIENCE	Not to take a shortcut
	To consult with task experts
	To consult with others

HONOR	To hold yourself and others to work safely even when no one is watching
LOYALTY	To the safety of your fellow associate
RESPECT	For the well-being of self and others
DISCIPLINE	To not prioritize safety
HONESTY	In safety participation
COMPASSION	For new hires and returning casualties

The visual model should also have an accompanying narrative that makes clear the dynamics and overall philosophy.

Next, culture must be defined by how it will be measured. As covered in chapter 1, culture has two levels; management associate and workforce associate. Each level has five elements to be measured. These measures become the overall measures for safety success that can be benchmarked to industry and to organizational goals and objectives. The five elements of culture are commitment, participation, perception, compliance, and competency. Table 2.5 below suggests one visual representation of organizing the measures of culture.

TABLE 2.5 *Five Elements of Culture*

LEVEL	ELEMENT	LEVEL
MANAGEMENT MEASURES		WORKFORCE MEASURES
Percentage of corrected safety problems from closed audits and project proposals.	Commitment	Percentage of Level 1 hazards self-identified and corrected.
Management representation and sign off on safety committees.	Participation	Percentage of accurate reports of incidents identified by workforce level associates of all known incidents.
Mean rating of perceptions from survey	Perception	Mean rating of perceptions from survey
Rating from ethical assessment of policy adherence	Compliance	Ratings from behavior based observations as compared with authorized and competent level scores
Percentage of management developed to safety competent level	Competency	Percentage of workforce developed to competent level

The management section builds on the strategic plan launched in the leadership section. It begins with long term and short term goals. Long term goals are tied to the vision statement, while short term goals improve on the strategy of the mission statement. All goals are supported by objectives, individual projects, or steps that will be completed in order to fulfill the goal. Objectives must be measureable. Objectives tell what specific activity will be completed, how and to what standard if applicable, and how it will be measured for success. Measurable means objectives are direct, statistically collected data; descriptive data; observable data; or evaluation scores. Keep in mind that level 1 and level 2 metrics can be correlated to level 3 performance indicators to indicate the level of effect on the final performance of safety. This concept guides us to base budget proposals and justifications on goals and objectives.

Goals are justified by the strategic plan of the organization and specific to safety from the safety operations plan (SOP) leadership section, relying on the vision statement. Objectives are justified by the guiding values and their projected impact on meeting the goal. Budgets then become justified by the strategic plan. Annual budget requests are justified and based upon the goals and objectives prepared for the upcoming year. Naturally, a miscellaneous budget account can be prepared for unforeseen safety needs.

Goals and objectives also indicate safety-program maturity. Programs mature, according to Kausek, from the design stage to the elimination stage. In the design stage, organizations create policy and measure success based on policy. In the compliance stage, policy is implemented and measures adhere to the policy. In the effectiveness stage, policy is measured for performance by examining level 3 or performance indicators. In the continual improvement stage, after policy is effective and measures examine or establish level 1 and 2 leading metrics, success is based on elimination of hazards (2009).

Akin to Kausek's model of safety program maturity, safety programs can have management endeavors that mature at varying rates. For example, suppose an organization is ready for behavior-based safety and begins its implementation well after implementing a hazard recognition initiative. Goals and objectives therefore can also mature year after year in the same way overall safety programs mature.

The following format in Figure 2.6 suggests one organizational structure for communicating goals and objectives in the SOP, along with one example.

LONG TERM GOAL # 1

XYZ Corporation will create and maintain a safety management system that meets the criteria set out in ISO 45001 international safety management system standard.

Justification: The vision statement indicates setting a level of performance exemplary to the industry. Exhibiting compliance to this standard furthers and indicates commitment to this vision.

Measure for Success: Score of compliance to the internal SMS audit based upon ISO 45001.

In the format above, a long term goal is to maintain a safety-management system that complies with a standard. The measure is indicative of a maturity level of design. However, other long term goals may be to maintain a high level of safety performance. Goals can combine to show maturity. An example of a short-term goal is given below.

SHORT TERM GOAL # 1

XYZ Corporation will reduce slips on permanent ladder surfaces by 50 percent as compared and projected from the experience of slips on permanent ladders in 2018 at the XXville Facility.

Objective 1: Install Slip-Not brand rungs on all permanent ladders in the XXville Facility.

Justification: The mission statement commits the organization to providing necessary resources for a safe and healthful work environment. In 2018, slips on permanent ladders comprised 50 percent of all slips, trips, or falls, totaling just over $18,000 in loss. Projected cost of installation with outside resources is $9,000. Project break-even point is six months.

Measure for Success: After installation is complete: Number of STF's involving permanent ladders at XXville Facility for 1 year.

FIGURE 2.6 An example of an organization structure for goals and objectives in an SOP.

The examples above reveal the structure for communicating short-term goals in the SOP and also reveal the strategy for budget justifications and show some of the information required for strategic project proposals. Line item budgeting and justifications where break-even points are calculated is a method called Zero Based Budgeting. Break-even points make sense for safety-related projects because returns on investment are sometimes difficult to project due to human error.

Proposing budgets based on Zero Based Budgeting is a constant duty for safety management. Goals and objectives can appear or change throughout the year of operations. Tracking for success can even go beyond fiscal years but is important in maintaining the credibility of the safety department to other divisions of the organization, as well as adjusting resource allocation for more effective safety performance.

The next piece of the management section of the SOP is to establish a format for project proposals with input from the accounting, purchasing or other relevant divisions of the organization. Audits, incident investigations, hazard-recognition programs, and any form of assessment may produce problems that require project proposals. Therefore, any investigation or activity that identifies a hazard, threat, or vulnerability that cannot be immediately corrected should be documented on a countermeasure report. If the correction is complex enough or identifies a correction that will require capital expenditure or approval from other cross-functional divisions, it will have to be proposed formally. The countermeasure report, whether created from safety or from a problem-solving committee, provides a perfect opportunity to measure allocation of resources and elimination or abatement of hazards. Remember that the ANSI hazard abatement strategy identifies corrective activity based upon the goal. Elimination is a change in task or role and a switch to less risky alternatives. Other types of controls, such as barrier guards, do not eliminate hazards; they balance exposure in an effort to prevent injury (ANSI B11.1, 2008).

A proper project proposal can be developed from combining an executive summary and a full audit report. An executive summary is a concise summary of information that executive management can read in order to get a five-minute version of the project at hand. The critical components of an executive summary begin with tracking formation, such as the incident control number, title of the project, dates, and who has proposed and initially approved the project details. The background information and justification is typically first. The justification is tied back to the goals and objectives of the SOP, and it presents the organization's experience with the problem. The experience includes expenditures and a description of the risk or probability, frequency, and exposure severity. Next is a brief of the project, broken down if necessary into project stages and assigned responsibility. The next section is the investment return. In the investment section the break-even point is calculated by comparing the required investment for needed resources versus the cost of similar incidents from the organization's experience. It would be a good idea to include a charted list of resources that reveals the overall needed resources, resources on hand, and required outside resources with cost. The final section of the executive summary should consist of follow-up dates and assigned responsibility for follow-up assessment. Figure 2.7 suggests one possible example of an executive summary.

Manufacturing LLC

Executive Summary

IC Number: _____   **Date of Report:** ___/___/_____

Approved By:_____   **Facility:** _____

To: Senior Management Team

Re: Installation of slip-resistant rungs on facility permanent ladders

Goal: Reduce work comp expenditures for slips/falls on permanent ladders.

Goal #1 from 2017 SOP: XYZ Corporation will reduce slips on permanent ladder surfaces by 50 percent as compared and projected from the experience of slips on permanent ladders in 2018 at the XXville Facility.

Justification: The mission statement commits the organization to providing necessary resources for a safe and healthful work environment. In 2018 slips on permanent ladders were causal in 50 % of all slips, trips, or falls, totaling just over 18,000 dollars in loss. Projected cost of installation with outside resources is $9,000.00. Project break-even point is 6 months.

Measure for Success: After installation is complete: Number of STF's involving permanent ladders at XXville Facility for 1 year.

Summary: In 2017 Slips, Trips, Falls accounted for $xxxxx.xx in expenditures. STF's are #1 in frequency and expenditure, justifying the targeting of this condition. Of these STF's, 6 occurred due to non-compliant surfacing on fixed industrial ladders. Risk rating: High

Project: Purchase Slip-Not Brand rung covers. Fabrication Section of Maintenance will install by welding: North end of plant by XX/XX/XXXX Responsibility: XXXXXXX Maint. Manager South end of plant by XX/XX/XXXX Responsibility: XXXXXXX Maint. Manager

Resources Required	Resources On-Hand	Needed Resources/Cost	Strategy

Break-Even Point: Total investment / projected savings per month
(Projected savings: Average cost of loss experience per month minus projected savings if investment meets goal per month.)

Projected Savings: 1 yr/_____ **3 yr/**_____ **5 yr/**_____

Follow-up: **Responsible Auditor:**

FIGURE 2.7 An example of an executive summary.

A complete report typically has four sections. The first section is the executive summary. The body of the report begins with the findings section. Here, the original causal analysis and any detailed narrative from the originating source is attached. The next piece of the report's body is the countermeasure report. The next piece of the proposal's body includes a break-down of necessary resources and how they will be obtained, quotes or required information for the cost of outside resources, a description of the project, estimated times of completion, estimates for contract work, details of any contractors, and a plan for managing contractor safety.

Core management programs are vital to develop and place in written form in the safety operations plan. Core programs differ from compliance programs in that core programs describe how services will be delivered to the customers or how safety will be managed. These vary from organization to organization and are not standardized from organization to organization because they are not based on compliance with a regulation or company policy. Core programs are based on organizational strategic plans and needs revealed by the needs of safety's customers. Many of these programs are mandated not only by organizational needs but also by safety management system standards. Typical core management programs begin with hazard recognition, investigations, emergency response, behavior based safety initiatives, and change management.

We will cover specific details of these programs in separate dedicated chapters in this book. However, the basic structure begins with an overall purpose, identifies any references, describes general requirements, assigns responsibility, details competency requirements, identifies training content, identifies measures of success, includes compliance level audits and program effectiveness audits, and includes any supporting process forms. This basic format will also be used to structure compliance programs. Compliance programs however, will take on a goal of being a teaching document as well as policy. Therefore, compliance programs will include schematics and how-to sections for completing commonly encountered situations.

REFLECTION 2.4

1. What other measures for culture do you envision?
2. How will you gather the data for your measures of safety specifically? Do supporting processes provide most of the specific data?
3. Why are goals and objectives so vital?

PROGRAM EFFECTIVENESS AUDITS

Each core program and compliance program will contain a program effectiveness audit for annual assessment of performance. This helps evaluate the entire safety program for continuous improvement efforts. Not only is this a requirement in all safety management system standards but it also allows the safety manager to adjust resource allocation to create the most positive effect with individual initiatives and the safety program in its entirety. Its goal then is to improve the program through individual initiative assessment.

The structure of our program effectiveness audit will mirror the Mathis structure of metrics and reflect the measures of culture (MOS). Level 1 metrics, or measures of management responsibilities, centered on resource allocation include training delivery and effectiveness, ratings of hazard correction by established prioritizing policy, and ratings of management's compliance to organizational ethics and policy. This first section is the management or "M" section.

The operational or "O" section covers measuring operational measures and culture perception. Operational measures, for example, include how well workforce associates follow established procedures. A specific example is an observation of authorized personnel conducting lock out tag out and scoring their adherence to company policy based on regulatory requirements. Other measures of compliance can also be included. Perceptions and attitudes can also be assessed in the operational section.

The status or "S" section includes the final performance measures that come from the individual initiative being assessed and from the performance indicators used to measure overall safety-program success. A typical example is the number of incidents attributed in cause to the individual initiative. Specific to lock-out tag-out (LOTO), we might report on the number of incidents known to be caused by improperly following or failing to follow lock out tag out. But status measures may also include the competency level of the various levels of responsibility assigned to the individual initiative.

The sections "M" and "O" are leading precursors to positive safety performance, while the "S" section is a lagging, final performance indicator. So measures of competency level as they stand now based on the levels of authorized, competent, or administrator are final performance measures. A measure of completed training percentage would be a level 1 precursor assessed under the "M" section because management has a duty to educate and train its associates.

The overall measurement criteria for the safety subset of organizational culture, commitment, participation, competency, compliance, and perception are also reflected in program-effectiveness auditing. It is important then to define specific measures or to give ranges and meaning to scores that the auditor or audit team will adhere to when assessing the established criteria. Table 2.6 reflects an example of defining audit criteria for assessing a compliance program of lock-out tag-out.

TABLE 2.6   *Management Level Criteria*

ITEM	CULTURE CRITERIA	CRITERIA	RESULT
1	Commitment	What is the percentage of production machines properly designed or equipped with proper lock out points and bleeding/blocking points? _____%	1 2 3 4 5

In the above example, management has a duty to provide the necessary equipment for following its procedure for LOTO. This criteria involves assessing the design of the production machinery itself. Other criteria can also measure commitment—issuing proper tags and locks to authorized personnel, for example. We could also assess how well management has communicated LOTO procedures to workforce associates, but this is one example. It is possible, for example, to have a machine with a pneumatic energy source or a compressed air line. That line would likely need a valve to bleed off air pressure for line repair. Merely shutting off the aid does not alleviate the hazard of compressed air still in the line. Suppose the organization reviews LOTO procedures on a schedule with a committee. Overall scores from these assessments can be used for this audit. Furthermore, we might set the expectations for purposes of scoring the results or findings. We may set the mark at 90 percent or above for a score of 5, or we may set 100 percent as meeting the requirements to score a 5. We could then scale down from that mark. The scale does not have to be even, either. We could say that below 90 percent is failing and corresponds to a 1. Typically, specific definitions for scoring and how to measure criteria may be covered in a separate page or manual for conducting the audit. But it is important to understand that effectiveness audits reflect cultural measures and may also rely on tallying the results of other audits, assessments, or investigations.

In measuring and reporting the success of the overall safety program of the organization, we should measure and report on culture through the 5 cultural criteria. Each initiative, or core and compliance programs, will be reviewed for effectiveness. The criteria used to assess their effectiveness will reflect culture. Once effectiveness reviews are completed for at least all core initiatives, the results can be used to assess overall culture of the entire program. This is in contrast to relying on final performance metrics to indicate safety success. Performance metrics are important, and when viewed in relation to quantified culture, reflect a more accurate summary of workplace safety at the organization.

REFLECTION 2.5

1. What criteria would you suggest for the Management Section of a LOTO program effectiveness audit?
2. What criteria would you suggest for the Cultural Section of a LOTO program?
3. What final performance metrics would you assess for a LOTO program?

CONCLUSION

Leadership and management combine to form a system that allows management to allocate safety resources accurately and efficiently while also leading all associates to accept less risk by moving safety from a priority in decision making to a virtue. The scope of the safety management system depends on the situation of the organization in its relationship of activity to the customers of the organization. Communicating the system to the customers in a written safety operations plan identifies the methods of providing those services and establishes a formal method for evaluating the system for positive performance.

REFLECTION 2.6

1. Why is organizational ethics so important to management?

REFERENCES

American National Standards Institute. *B11: 2008 General Safety Requirements Common to ANSI Machines*. McLean, VA: The Association for Manufacturing Technology, 2008.

Crossan, M. M., M. J. Rouse, J. N. Fry, and J. Killing. *Strategic Analysis and Action*, 7th ed. Toronto: Prentice Hall, 2009.

Heinrich, H. W., Dan Peterson, and Roos Nestor. *Industrial Accident Prevention*. New York: McGraw-Hill, 1980.

Kausek, J. *OHSAS 18001: Designing and Implementing an Effective Health and Safety Management System*. Lanham, MD: Government Institutes, 2007.

Mathis, Terry L. "Common Practice: The Third Level of Leading Indicators." *EHS Today* (2014). ehstoday.com/safety-leadership/common-practice-third-level-leading-indicators.

Manuele, Fred, A. *Heinrich Revisited: Truisms or Myths*. Chicago; National Safety Council Press, 2002.

Figure Credits

CHAPTER THREE

Conducting the Hazard Inventory

FOREWORD

To manage safety, you have to know what you are dealing with. The first step to successful safety management begins with identifying the general hazards present in the facility. This inventory of safety topics gives an overhead view of what must be managed. Depending on the maturity of the organization, there may be resources in existence that can be referenced prior to any physical inspection of the activities, machinery, processes, or facility. Official job descriptions, job hazard analyses, worker's compensation records, OSHA-required injury logs, and industrial classification numbers are good resources to begin understanding what overall hazards are present. But these are supplemental to a physical walk-through assessment. The hazard inventory will serve as the overall guide for the safety manager to understand what regulations are central for basic compliance.

Hazard recognition for workforce associates is different than hazard recognition for safety professionals. The workforce associate must recognize hazards from the standpoint of a category leading to a specific source. For example, workforce associates should be able to identify an impact hazard, or something that has potential to strike them and the source, such as a forklift traveling in an aisle. Safety professionals must be able to recognize hazards as well, but also be able to cross reference the categories to regulatory concern. This begins with understanding activity and identifying appropriate regulation.

Objectives
After reading this chapter, the learner will be able to do the following:

1. Identify legal liability to specific regulations contained in the Code of Federal Regulations in reference to the proper part of CFR based on industry activity

2. Identify and formulate an inventory of specific legal liabilities presented by industry hazards and activities according to subparts of 29 CFR Part 1910
3. Perform an emergency event inventory for planning purposes
4. Perform an inventory for vulnerabilities as a prelude to security planning

LEARNING PLAN

LEVEL	ULTIMATE OUTCOME	CLAIM	LEARNING TYPE	ASSESSMENT
F	Identify legal liability to specific regulations contained in the Code of Federal Regulations in reference to the proper part of CFR based on industry activity	Managers of workplace safety must be able to identify the proper legal requirements based on industry activity.	CR/AC	
U	Walk through a part of CFR to identify and formulate an inventory of specific legal liabilities presented by industry hazards and activities	Workplace safety management begins with understanding the legal liability of the organization in relation to its activity.	IL/AC	Learners will develop a hazard inventory for assessment of legal liability based on a part of the Code of Federal Regulations.
U	Perform an emergency event inventory for planning purposes	For proper emergency response planning to take place, a manager of workplace safety must inventory for emergency events.	CA/AT	Based on a fictitious facility and location, the learner will perform an inventory for emergency events.
U	Perform an inventory for vulnerabilities as a prelude to security planning	For proper security planning to take place, a manager of workplace safety must inventory for vulnerabilities.	CT/AT	Based on a fictitious facility (or known volunteer facility) and location, the learner will perform an inventory of vulnerabilities.

Learning Plan Legend

Level:

F: <u>Foundational outcomes</u>: Basic abilities

M: <u>Mediating outcomes</u>: Progress through a developmental model; interpret, analyze, evaluate progressively challenging claims, arguments

U: <u>Ultimate outcome</u>: Navigate most advanced arguments and claims

Type of learning:

CR: <u>Critical reading</u>: The ability to read, process, and understand the meaning of written information

IL: <u>Information literacy</u>: Locating and selecting suitable information for a task; evaluating appropriateness/validity of information sources

AC: <u>Application of concepts</u>: Ability to apply discipline-specific knowledge/skill to tasks/situations important to the discipline

CT: <u>Critical thinking</u>: Ability to apply a concept to a vague or argumentative claim without a creative leap

AT: <u>Analytical thinking</u>: Ability to critique/analyze situations using a concept or model

CA: <u>Creative application</u>: Ability to apply a model/concept in a new way/to an unrelated situation or scenario; involves creative leaps

DISCOVERING, DOCUMENTING, AND MAPPING REGULATORY SAFETY CONCERNS

The North American Industry Classification System (NAICS) is a coding system developed to categorize industry. By using this numbering system, statistical data can be compiled for industries that use similar raw materials, equipment, or labor activities. The numeric system classifies all industry by two to six digits. As more numbers are added, the specific activity of the industry is narrowed. Safety managers can access data from the Bureau of Labor Statistics to compare injury rates and types of incidents and identify trends that uncover general categories of hazard.

Some examples of industry classification are textile mills (NAICS 313), chemical manufacturing (NAICS 325), or primary metal manufacturing (NAICS 331). The classification alone gives guidance to some general hazards and provides an avenue to research for safety concerns. Textile facilities deal with certain dusts that can cause byssinosis. Inhaling cotton, flax, or hemp particles can lead to a serious lung condition, commonly called brown lung disease. Information like this may be common knowledge to some, but the classification of industry can aid in researching safety issues. One primary resource is the Bureau of Labor Statistics. The "industry at a glance" feature gives quick and reliable overviews.

The codes of federal regulations for workplace safety cover various industries by activity under Title 29 of the Code of Federal Regulations (CFR). Different parts of the CFR apply to more specific activities. Activity on waterways within US borders are covered by maritime regulations contained in Part 1917, transportation regulations are covered under Part 1940, construction is covered under Part 1926, Part 1915 covers shipyard employment, Part 1916 covers long-shoring and marine terminals, and general industry is covered under Part 1910. In this chapter we will walk through Part 1910 in a manner that helps us identify specific legal liability for a general industry facility. The same technique would be applicable to other parts of CFR, such as 1926.

General industry regulations under 1910 can apply to public employees in "state" plan states or in states that have passed legislation placing public employees under federal OSHA regulations. That means in states such as Kentucky, even police departments, fire departments, and schools are subject to state OSHA regulations. While state regulations are largely adopted federal regulations, they can be more restrictive.

Once the applicable part of the CFR is identified, we can begin to look at the subparts and specific standards that reveal general safety concerns for inventory purposes. A look at the overall table of contents is applicable. For example, when we look at the contents of Part 1910 of 29 CFR, we see that walking and working surfaces is the first specific safety concern covered by regulations. In Section 22 the general requirements for walking and working surfaces are covered. Since all places of employment deal with walking and working surfaces, and since this is also a category of hazard that has multiple sources, we can break down the safety concerns of the regulations in a way that helps us inventory what we are specifically concerned about in the facility.

Breaking down Subpart D of 29 CFR 1910 we can begin to note specific safety concerns. With housekeeping in regard to walking and working surfaces we see that clean and orderly is a universal requirement. Additionally, from looking at Subsection 2 of Paragraph a, wet processes or areas where wetness is common is an area to be inventoried. Aisles and passageways including permanent aisles are other areas to be inventoried. Next we would need to identify all walking and working surfaces near special hazards such as drains, ditches, or pits. Then we would need to identify separate floors for loading requirements. The idea is to identify the presence and location of such safety concerns from a regulatory standpoint and then examine for general categories of hazard to be as thorough as possible. Table 3.1 displays the inventory strategy and format for walking and working surfaces.

TABLE 3.1 *Subpart D Inventory Strategy*

REGULATION	CONCERN	LOCATION	HAZARD	SOURCE
W/W Surfaces 1910.22 Housekeeping	Permanent aisles			
W/W Surfaces 1910.22 Housekeeping	Pits, ditches, drains, vats, special hazards			
W/W Surfaces 1910.22 Housekeeping	Floors			

The chart displays the special concerns for walking and working surfaces in Section 22, housekeeping. These should also be identified on a map or schematic of the facility. We are not yet creating specific compliance level audits for all the paragraphs in Section 22, but are identifying the areas where this audit will be applicable. Upon physical inspection of these areas, we can identify any categories of specific sources and hazards found.

Table 3.2 shows the concerns for the entire subpart for walking and working surfaces. This type of inventory would have to be completed for all subparts of 1910 for general industry application.

TABLE 3.2 *Subpart D Inventory*

REGULATION	CONCERN	LOCATION	HAZARD	SOURCE
W/W Surfaces 1910.22 Housekeeping	Permanent aisles			
W/W Surfaces 1910.22 Housekeeping	Pits, ditches, drains, vats, special hazards			
W/W Surfaces 1910.22 Housekeeping	Floors			
W/W Surfaces 1910.23 Floor/Wall Holes	Wall openings/ holes			

Continued

REGULATION	CONCERN	LOCATION	HAZARD	SOURCE
W/W Surfaces 1910.23 Floor/Wall Holes	Open side floors, platforms, runways			
W/W Surfaces 1910.23 Floor/Wall Holes	Stairways			
W/W Surfaces 1910.24 Fixed Stairs	Fixed stairs			
W/W Surfaces 1910.25-1910.26 Portable Wood/Metal Ladders	Wood ladders, metal ladders			
W/W Surfaces 1910.27 Fixed Ladders	Fixed ladders			
W/W Surfaces 1910.28 Scaffolding	Scaffold type			
W/W Surfaces 1910.29 Mobile Scaffolding/ Ladders	Mobile ladders, scaffolds			
W/W Surfaces 1910.30 Other Surfaces	Dockboards			
W/W Surfaces 1910.30 Other Surfaces	Forging machine area			
W/W Surfaces 1910.30 Other Surfaces	Veneer machinery			

A facility schematic that maps out locations of the specific concerns is very useful in the planning aspect of safety management. There are other concerns that may not be referenced in regulations as well. Outdoor walkways or tiled floors in kitchen areas are examples that you may encounter and want to identify in the initial inventory for walking and working surfaces. The same can be said for concerns in any subpart.

The order of 1910 will next cover exit routing and emergency plans in Subpart E. Sections 33 through 39 cover this topic. Regulations point to the need to identify several areas of concern. These areas include exit routes, exits, exit signage, emergency lighting, alarms, outside evacuation routes, muster areas, posting areas, major fire hazards, ignition sources, fire protection equipment, heat producing equipment, and a listing of all hazardous materials, which is also a part of hazardous material inventory. Experience in managing emergency scenes and planning identifies other concerns that require initial mapping and inventory. These include points of entry and utility routes for incoming utilities such as water, electricity, and natural gas, and their corresponding shut-off mechanisms. Other examples include the following:

- Facility response routes from first responder stations
- Secondary access routes for first responders
- Helicopter landing area grid coordinates
- Post indicator valve and hydrant locations
- Standpipe locations and controls
- Hazardous material storage rooms and other storage piles
- Emergency equipment storage areas

Subpart F covers powered platforms, manlifts, and vehicle-mounted platforms. This subpart is more specific to certain types of equipment. Inventory of powered platforms, manlifts, and vehicle-mounted systems will be useful. In this inventory it is helpful to include the make, model number, and manufacturer of the equipment. Inventorying the entire list of heavy and powered equipment helps in identifying the regulations and special concerns you will deal with in your organization. One special concern is for recall notices. The safety manager is typically responsible for verifying that any manufacturer-ordered repair or alteration is completed as specified by the maker and for signing a certificate guaranteeing that the repair was completed. Sometimes a repair or modification may have to be completed by an authorized vendor, but the safety manager must still maintain documentation and inspect for the continuity of the repair. One simple example might be to install a specific warning sign at a specified location on the machine. Of course, this should have a duplicate tracking mechanism through the maintenance department. The maintenance department documents all histories of repair, alteration, and problems with machinery to include warranty and recall history. Therefore, the maintenance department is a valuable source for information pertaining to equipment inventory. Table 3.3 displays a possible template for equipment inventory purposes.

TABLE 3.3   *Subpart F Inventory*

EQUIPMENT DATA	REGULATORY STANDARDS	SPECIAL CONCERNS
Make:		Recall dates:
Year:		
Model#:		
Serial#:		
Type/description:		
Fuel:		

Subpart G covers occupational health and environmental control issues such as ventilation and noise. Here, the safety manager must identify specific operations where ventilation is of issue and perform a noise level assessment. Section 94 covers abrasive blasting operations for ventilation, as well as grinding, polishing, and buffing operations and spray finishing operations. Ventilation is an issue around coating operations, to include many dip processes, around sources of heat, and any dust producing operation such as welding. Noise surveys are often an offered service of worker's compensation insurance providers. In this assessment, noise levels are measured throughout the facility in a way that establishes boundaries. Mapping these boundaries helps predict individual exposure but allows the safety manager to clearly mark and define areas of high noise exposure above the action threshold or higher levels to better manage personal hearing protection and enforcement and identify areas where more engineering controls are warranted or areas where administrative management limits access.

Subpart H is concerned with hazardous materials. The first step in hazardous material inventory is to list each individual hazardous material, where it is located for use and storage, and identify the least amount of stored material necessary for continued production. By using material safety data sheets and management tools such as CAMEO (computer-aided management of emergency operations), a software hazardous material management program, downloaded free from the National Oceanographic and Atmospheric Administration or from the Environmental Protection Agency, the safety manager can determine what hazardous materials are compatible for storage as well as learn more about specific handling, storage, and emergency exposure and response procedures.

Subpart H is specifically concerned about the following items contained in Table 3.4.

TABLE 3.4 *Subpart H Inventory*

PART/SECTION	TOPIC/MATERIAL	APPLICATION	LOCATION, IF APPLICABLE
1910.101	Compressed gas MSD present Yes ☐ No ☐		
1910.102	Acetylene MSD present Yes ☐ No ☐	Cylinders: Piped systems: Generator/filling:	
1910.103	Hydrogen MSD present Yes ☐ No ☐	Gaseous: Liquefied:	
1910.104	Oxygen MSD present Yes ☐ No ☐	Bulk systems: Containers: Piping: Vaporizers:	
1910.105	Nitric oxide MSD present Yes ☐ No ☐	Piped: Containers:	
1910.106	Flammable liquids MSD present Yes ☐ No ☐	Liquid: Flammability cat: Container: Storage area: Piping:	
1910.107	Spray finishing using flammable/combustible materials	Material: MSD present Yes ☐ No ☐ Operation:	
1910.109	Explosives/blasting agents	Material: Class: MSD present Yes ☐ No ☐ Process: Storage: Packing/transport:	
1910.110	Liquefied petroleum gas	Container: MSD present Yes ☐ No ☐ Amount per container: Number on hand: Piping: Vaporizer: Filling equipment: Storage:	

Continued

PART/SECTION	TOPIC/MATERIAL	APPLICATION	LOCATION, IF APPLICABLE
1910.111	Anhydrous ammonia	Container: MSD present Yes ☐ No ☐ Piping: Transfer area: Storage: Refrigerated: Stationary: Tanks: Process:	
1910.112	Process	Chemical: MSD present Yes ☐ No ☐	
1910.120	Hazardous waste operation	Material: MSD present Yes ☐ No ☐ Lab analysis Yes ☐ No ☐ Process: Waste stream: EPA class: Listed: Characteristic: Universal: Used oil: DOT class: Storage:	
1910.123- 1910.126	Dipping and coating operations	Use: Clean ☐ Coat ☐ Alter ☐ Material: MSD present Yes ☐ No ☐ Operation type:	

In this inventory list it is important to note that general categories such as "flammable liquids" should have the information inventoried for each flammable liquid at the facility. These include spray operations, explosive and blasting agents, processes, and hazardous wastes. Hazardous wastes are critical for environmental aspect inventory. An "aspect" is an environmental hazard. Environmental management begins with aspect inventory mirroring the hazard inventory. Aspect inventory includes any source and material that poses a risk to plants, animals, humans, fish, or other organisms through air, water, and soil contamination.

Subpart I covers personal protective equipment (PPE) and has a special requirement under 1910.132(d) mandating that the employer assess the workplace for hazards that are present, which necessitates and requires use of effective personal protective equipment, communicating it use to employees, and providing properly fitting options. This PPE assessment should be completed after the hazard inventory. The hazard inventory can produce a list of hazards, processes, machinery, and areas to assess. Table 3.5 suggests a possible inventory assessment instrument for PPE.

TABLE 3.5   *Subpart I Inventory*

Machine:	Hazards:	PPE category:	Type:
Process:	Category:	Hand ☐	Specifications:
Task:	Source:	Head ☐	
Location:	Exposure level:	Ear ☐	
Job:		Eye ☐	
		Foot ☐	
		Body ☐	
		Respiratory ☐	

The safety professional can identify the machine, process, task, job, or location of special area that will require personal protective equipment by using Table 3.5. The next step is to identify the specific hazards to be addressed by category and then by source. For example, it may be a chemical hazard with a specific source or chemical of nitric acid. The evaluator can then check all PPE categories that will be required at this location and then for each type of protective gear list the specifications that match the hazard. The example might reflect ear plug and specifically the specific type of ear plug and noise reduction rating level. The safety professional can complete this template for as many categories of different PPE that are necessitated by the hazards presented. This assessment template will apply to only one category of PPE at each machine. It could be modified to represent multiple categories.

Subpart J covers general environmental controls. Water supplies, waste disposal, vermin, toilets, break areas, and signage are important for inventory. Some of these will also be valuable for other areas of management and planning such as locations of wastes for fire prevention planning. Table 3.6 represents important inventory topics for this subpart.

TABLE 3.6 *Subpart J Inventory Topics*

PART/SECTION	TOPIC/MATERIAL	APPLICATION	LOCATION, IF APPLICABLE
Subpart J 1910.141	Sanitation	Waste collection points	
Subpart J 1910.141	Sanitation	Signs of vermin	
Subpart J 1910.141	Sanitation	Water supply, potable	
Subpart J 1910.141	Sanitation	Water supply, non-potable	
Subpart J 1910.141	Sanitation	Toilets	Location: Type: Number: Location: Type: Number Location: Type: Number
Subpart J 1910.141	Sanitation	Washing facility type:	
Subpart J 1910.141	Sanitation	Showers	
Subpart J 1910.141	Sanitation	Change room	
Subpart J 1910.141	Sanitation	Clothes drying points	
Subpart J 1910.141	Sanitation	Uniform collection points	
Subpart J 1910.141	Sanitation	Food consumption/preparation area	

Subpart J 1910.142	Temporary labor camp	Camp locations	
Subpart J 1910.144–145	Safety signage	Safety signage/markings	Type: Location: Type: Location: Type: Location:
Subpart J 1910.144–145	Safety signage	Communication posting/ board areas	

Subpart J also contains confined spaces and lock-out tag-out requirements. Confined spaces are defined by OSHA as having three elements: (1) large enough and configured so that a worker can bodily enter and perform work, (2) limited or restricted means for entry or exit, and (3) not designed for continuous employee occupancy (1910.146(b)). Common examples include tanks, pits, and silos, but you must also consider platforms or access areas to specialized equipment or at height, since entry and exit also include rescue or access for treatment. A permit-required confined space is a confined space that contains any recognized serious safety or health hazard. These include, by definition, a hazardous atmosphere or potential to have a hazardous atmosphere, a potential for engulfment, or a special internal configuration that could trap or cause asphyxiation (1910.146(b)). It is vital that inventory identify, mark, and map these areas.

Lock-out, tag-out (LOTO) presents a slightly different challenge to inventory. Basic inventory should identify locations of special lock-out equipment that is stored for multiple authorized persons to access as needed. Examples of special LOTO equipment might include group lock-out boxes, special lock-out devices such as gate valve locks, plug covers, circuit breaker locks, and many other special devices. LOTO program management requires most every machine, process, or piece of equipment to have developed and documented procedures for rendering it at a zero-energy state. These are usually mapped at the specific location of use identifying the hazardous energy type and source; have pictures of the lock-out location, bleed or blocking points; and list specific steps listed in sequential order for rendering the item at zero energy. But this is a process not of inventory. Table 3.7 reflects inventory for Subpart J and confined or permit-required confined spaces.

TABLE 3.7 *Subpart J Inventory*

PART/SECTION	TOPIC/MATERIAL	APPLICATION	LOCATION, IF APPLICABLE
Subpart J 1910.146	Permit-required confined spaces	Confined spaces	Description: Location:
			Description: Location:
			Description: Location:
			Description: Location:
Subpart J 1910.146	Permit-required confined spaces	Permit-required confined spaces	Description: Location:
			Description: Location:
			Description: Location:
			Description: Location:

Subpart K covers medical services and first aid. Here, inventory is vital for emergency-response planning. Managing even first aid-level response requires considerable inventory of locations. First-aid boxes are an obvious concern for inventory. However, there are two levels of first-aid response in practice. It is irresponsible to provide access to equipment that employees have not had proper training or authority to use. Secondly, 1903.14(f) places liability on an employer that does not properly delineate rescue duties. Therefore, basic first-aid boxes must contain only items that those with access can use competently. If the boxes are to be accessible by all, it is probably too resource expensive to place trauma-level items in the box. These items might include CPR masks, blood clotting substances, or advanced medicines that require documenting nonprescription usage amounts. So, in locations of controlled access to those trained and certified by the company for rescue and use of trauma-level equipment, such as CPR masks or defibrillators, more advanced tools and equipment must be staged for higher-level emergencies. Inventory of these locations is initially necessary to facilitate proper response planning that meets the reasonable response time requirement of four minutes established in the Brogan Letter of Interpretation.

Although not mentioned directly in the regulation, in many cases emergency equipment such as eyewash stations, showers, and defibrillators are connected to alarms. Inside many facilities, radios or phones are used to summon first-aid assistance due to announcing the location. Table 3.8 represents inventory needs for medical and first-aid program management.

TABLE 3.8 *Subpart K Inventory*

PART/SECTION	TOPIC/MATERIAL	APPLICATION	LOCATION, IF APPLICABLE
1910.151	Medical services/first aid	First-aid boxes	
1910.151	Medical services/first aid	Trauma equipment bo	
1910.151	Medical services/first aid	Phones for summoni applicable alarms	
1910.151	Medical services/first aid	Nearest hospital, station, nearest	
1910.151	Medical services/first aid	Eye wash stati shower	

Fire protection is the subject for subpart requirements for fire brigades, portable and fixed fire equipment, n systems, and fire- or employee-activated alarm systems. Rather t aking inventory of what and where, it would serve long-term plann to not only inventory for these systems and components, but to inve hazards as well. The safety professional can cross-reference the existence of a rd to a system that is appropriate for the hazard. Inventory applicable for subpart L will also include a facility map locating fire extinguishers, fixed systems identified such as standpipes, post indicator valves, sprinkler risers, foam systems, dry chemical agent–fixed systems, gas agent systems, fire detection systems, and employee-activated alarms. The fire protection map should also denote any flammable stores, hazardous material storage rooms, and any special features of the facility such as blow-out walls or areas of special fire protection. Table 3.9 is one such suggestion for an inventory instrument for fire hazards and Subpart L. Each hazard or system should have a documented inventory of the basic included information on the inventory instrument.

TABLE 3.9 *Subpart L Inventory*

FIRE HAZARD NOTES	NAME	LOCATION/ STORAGE	PROTECTION SYSTEM	SPECIFIC
Flammable liquid ☐	Common: _____	Location:	Portable: ☐	
Flammable aerosol ☐	CAS: _____		Fixed: ☐	
Combustible pile ☐	UN: _____	How stored:	Standpipes: ☐	
Flammable store ☐	DOT class:	_____	Automated	
Electrical fire ☐		Container: ☐	Sprinkler: ☐	
Combustible metal ☐	Task/process:	_____	Foam: ☐	
		Tank: ☐	Dry chemical agent: ☐	
		Pressurized tank: ☐	Agent: _____	
		Piping:	Gas agent: _____	
		Drum: ☐		
		Drum size:	Fire detection: ☐	
		Drum type:	Employee alarm: ☐	

Subpart M is rather short and is applicable to compressed air and gas equipment. With this section, inventory covers the mapping and inventory of all air equipment, compressed gas systems, and what machines or areas each plant system covers. This inventory will be valuable in planning lock-out tag-out procedures for the systems and individual machines, reporting requirements for hazardous materials in the form of compressed gases, and in planning emergency shutdown duties. Table 3.10 suggests the inventory instrument for compressed air/gas systems.

TABLE 3.10 *Subpart M Inventory*

PART/SECTION	SYSTEM/TYPE	LOCATION(S)	MACHINES/PROCESS
1910.169 Compressed Air/Gas Systems	Pneumatic: ☐ Compressed gas: ☐ Gas agent: _____		
1910.169 Compressed Air/Gas Systems			
1910.169 Compressed Air/Gas Systems			

With this inventory instrument the inspector can identify whether the system compresses aid or delivers a compressed gas. The location of the tank, compressor, and piping can be mapped on a facility map and documented along with the machines or processes it serves. For example, in a large fabrication shop, many welding machines may be served by a large compressed gas delivery system of Argon. The system may have a large storage tank outside the plant and deliver shielding gas to specific operations. Another example might be a compressed air system that delivers compressed air from a large air compressor to many power presses on the plant floor.

Subpart N covers material handling and storage. Specifically, forklifts cranes and derricks are covered and of interest to the inventory. However, other issues surrounding these pieces of equipment are also important, such as storage rack types, clearances, and load. Mapping is also important and integral in material handling inventory. The widths, overhead clearances, permanent widths, and even important intersections of travel, passing areas, or other features of floor layout that encourage efficient flow of material-handling equipment and pedestrian traffic is important. This is an appropriate point at which to examine the materials stored on racks and the height of storage. This is important when assessing if the material would require rack sprinkler systems or if the ceiling sprinkler system is adequate for the hazard. Table 3.11 suggests inventory particulars covering this area that will aid in planning and managing material handling and storage.

TABLE 3.11   *Subpart N Inventory*

PART/SECTION	TOPIC/MATERIAL	APPLICATION/LOCATION	SPECIFICATIONS
1910.176	Handling materials	Aisle:	Width: Clearance:
		Loading dock:	Width: Clearance: Drop height: Special equipment: ———
		Doorways:	Width: Clearance:
		Permanent aisle:	Width: Clearance:

Continued

PART/ SECTION	TOPIC/MATERIAL	APPLICATION/LOCATION	SPECIFICATIONS
1910.176	Handling materials	Rack storage: Location: Capacity: Aisle width:	Location: Material: Combustible range: Sprinklers: Rack: Ceiling: Head size: Height to head:
1910.151	Handling materials	Railroad dock	Bumper blocks Yes ☐ No ☐
1910.177	Servicing material handling equipment	Maintenance area	
1910.177	Servicing material handling equipment	Cage for split rim wheels	
1910.178	Powered industrial trucks	Forklift type (CIRCLE) D, DS, DY, E, ES,EE,EX, G, GS, LP, LPS Location used: Charge station:	Make: Model: Lift capacity: Weight: Tires:
1910.179	Cranes	Type: Gantry: ☐ Overhead: ☐ Wall: ☐ Location:	Make: Model: Capacity: Controls: Floor: ☐ Cab: ☐ Remote: ☐
1910.180	Mobile cranes	Truck: Crawler: Locomotive:	Make: Model: Capacity:

PART/ SECTION	TOPIC/MATERIAL	APPLICATION/LOCATION	SPECIFICATIONS
1910.181	Derricks	Type: A-frame: ☐ Basket:: ☐ Breast:: ☐ Chicago: ☐ Gin: ☐ Guy: ☐ Stiff leg: ☐	Make: Capacity: Hoist: Manual: ☐ Powered: ☐
1910.183	Helicopters	Location: Owner:	Make: Model: Capacity:
1910.184	Slings	Type: Chain: ☐ Metal mesh: ☐ Wire rope: ☐ Synthetic: ☐ Locations:	Maker: Vertical cap: Softeners: Yes ☐ No ☐ Crescents: Yes ☐ No ☐ Length: Size: Rope: Lay: Twist:

Machine guarding is the subject for Subpart O. Machine guarding is a critical component of occupational safety. Inventory consists of identification of the type of machinery and the type of operator safety device utilized. However, for more advanced planning and integration of other safety requirements such as lock-out tag-out, or control of hazardous energy, it makes sense to inventory each separate machine for energy sources, bleeding or blocking points, and personal protective equipment. Usually, each machine or workstation will have an administrative area where safety and quality postings are located. Safety must at a minimum post lock-out tag-out points and procedures, list required personal protective equipment, and include a job hazard analysis. While inventory for each machine is conducted, it only makes sense to check for these minimal postings. Each machine or workstation present should be inventoried for the basic information included in Table 3.12. The form can be manipulated for the specific types of machinery present at the facility or site.

TABLE 3.12 *Subpart O Inventory*

MACHINE	SAFETY DEVICE	ENERGY SOURCE	ENERGY CONTROLS	POSTINGS	SPECIAL TOOLS
ID: Type: Woodworking ☐ Abrasive wheel ☐ Mills ☐ Power press ☐ Cutter ☐ Shear ☐ Forming ☐ Rollers ☐ Uncoiler ☐ Welder: ☐ Other: ☐	Operator: Two-hand control ☐ Light curtain ☐ Other: _____	Electric: Pneumatic: Water: Hydraulic: Compressed gas: Fuels: Kinetic: Other:	Identified at machine for all energy sources? Shut off: Yes ☐ No ☐ Bleed Point: Yes ☐ No ☐ Block Point: Yes ☐ No ☐	Located at workstation? PPE: Yes ☐ No ☐ LOTO: Yes ☐ No ☐ JHA: Yes ☐ No ☐	Name of tool: Type: Present: Yes ☐ No ☐

Once the safety professional begins to plan specifically, many production machines will require more specific information. Table 3.12 provides a basic picture of what must be dealt with for regulatory requirements in general. This inventory instrument can be enhanced with additional information that is useful in individual management initiatives and is specific to special machines. For power presses, as an example, it should include a complete description such as straight side or C frame and drive mechanism; manufacturer and features of the press controls; a full description of operator controls to include modes of operation, parts or products produced, and specific adjustments to the press; operations procedures; stopping times for each product application; lock-out tag-out procedures and times; die change procedures and times; and all safety features or controls present. The data listed impact the safety of press operations and are critical for safety professionals to be familiar with to properly coordinate safety with operators, supervisors, engineers, maintenance, and die setters.

Subpart Q, welding, cutting, and brazing presents a different challenge in inventory as compared to other subparts. Here, we may have identified some hazardous materials or process previously. However, being redundant here may prevent overlooking chemical hazards present in weld wire, flux, shielding gases, steel, or any special coatings on the welded materials.

First we must identify and map the type of welding taking place at the workstation or area and then the materials involved. Arc welding refers to a fusion process where heat melts the metal surfaces to be bonded and intermixes with a filler metal and as the new mix cools and hardens in a metallurgical bond. Typically this involves a stick of metal filler or a wire (mig welding) that contains the filler and may also contain a covering that shields the molten pool from contaminants in the air. Contaminants can weaken the bond. These shielding materials can be shielding gases emitted at the end of the mig gun or be solid coverings on a weld stick. Coverings or coatings that shield are referred to as flux. Resistance welding or ERW are used in spot and seam welding processes. Here, the heat to weld comes from the resistance of the metal material to be bonded, which takes time and pressure to bond the materials together. Figure 3.1 shows arc welding from a mig process.

Figure 3.2 shows the arc welding process from a stick process.

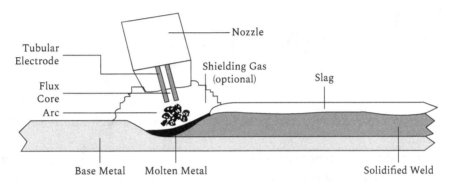

FIGURE 3.1   Flex Cored Arc Welding (FCAW) drawing

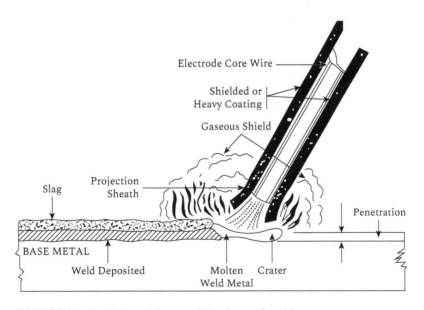

FIGURE 3.2   Shielded Medal Arc Welding (SMAW) weld area

FIGURE 3.3   Puntlasmachine

These two processes are common examples of welding processes in industry. Another process common in manufacturing is electric resistance welding. In ERW processes, the heat builds up from the resistance of the metal and creates a spot weld. Figure 3.3 shows a common foot-operated spot welder. The plates or metal surfaces get compressed between the two weld tips.

These are not all of the welding processes; for example there can be a submerged process, submersion in powder flux, or submersion beneath a liquid coolant. But the point is to identify the process and materials because this reveals much about the hazards encountered.

For each workstation that is a welding process or portable welding machine, we can gather initial information for hazard inventory purposes. Table 3.13 reveals the inventory requirements for managing safety issues with subpart Q.

TABLE 3.13   *Subpart Q Inventory*

TYPE OF WELD PROCESS	LOCATION/IDENTIFICATION	MATERIALS
ARC: MIG TIG stick plasma	Plant locale:	Metals:
	Workstation ID:	Wire:
ERW:		Stick:
Non-fusion welding:		
Solder:		Gas:
Braze:		Location stored: Amount: Type of storage:
		Flux:

Subpart R covers special industries. These industries are pulp, paper, and paperboard mills, textiles, bakery equipment, laundry machine operations, sawmills, logging, telecommunications, electric power, generation and distribution, and grain handling facilities. If the facility has any of this equipment or engages in these activities, identifying and mapping these activities and relevant equipment would be part of the hazard inventory.

Subpart S covers electrical safety in four areas: design for electrical utilization systems, safety-related work practices, maintenance requirements, and special equipment. The main concern in the hazard inventory is to identify and map where the utility enters the property and facility (which may be multiple points), where electrical rooms and other control points are located, and where special equipment and any electrical controls to special equipment that require special shutdown procedures in case of emergency are located. In regard to cabinets, boxes, panels and other control or junction points, it would be beneficial to also identify the voltage, which should be clearly posted. This will assist in arc-flash assessment. Special systems of 600 volts or more should also be located and mapped for special requirements under 1910.308.

Special equipment that would need to be located and mapped includes cranes and hoists, elevators, welders, X-ray equipment, heating equipment, electrolytic cells used in the production of aluminum, cadmium, chlorine, copper, fluorine, hydrogen peroxide, magnesium, sodium, sodium chlorate, and zinc. Any electrical equipment associated with water such as irrigation equipment or large water pools and tanks, or electrical equipment in hazardous locations or areas where flammable or combustible concentrations of gas or dust may collect, should also be located and mapped.

Table 3.14 suggests one such example for completing the hazard inventory for subpart S.

TABLE 3.14 *Subpart S Inventory*

LOCATION/DESCRIPTION	VOLTAGE	METHOD
Right of way/ property		Overhead lines ☐ Underground ☐
Electrical rooms		
Main cabinets/boxes, junctions		
Hazardous location Gas ☐ Dust ☐		Substance(s)

Subpart Z deals with toxic and hazardous substances. Most of these were probably inventoried previously in the hazardous materials section and may also be listed in an environmental aspect inventory. Environmental management begins with knowing environmental hazards or aspects, so an inventory similar to a hazard inventory is completed. This aspect inventory can also be used to ensure that substances are not overlooked in the hazard inventory.

Specific to Subpart Z is an inventory for air contaminants or carcinogens that can enter the body through mucous membranes such as the eyes, nose, mouth, or through intact or open skin. These may include total dust, lead, asbestos, chromium VI, cadmium, or body fluids. A comprehensive list is found in Table Z-1 of 29 CFR 1910.1000 and in the appendix to this chapter.

In Subpart Z, blood-borne pathogens are covered. Particular to this program, the safety professional who is performing a hazard inventory must assess and identify any associates who have occupational exposure. Occupational exposure is defined as "reasonably anticipated skin, eye, mucous membrane, or parenteral contact with blood or other potentially infectious materials from the performance of an employee's duties" (29 CFR 1910.1030). This would include any associate who is assigned first-aid response or rescue duties.

In addition to employees with occupational exposure, it would be careless not to recognize the reasonable potential for associates to have allergies and conditions such as diabetes that may require them to bring syringes onto company grounds. Without covering best practices for this situation, having a program to control needle-stick injuries is valid and necessary. Associates bring with them risk that is personal but extends into the workplace. These personal conditions may necessitate safety policy and procedures that control exposure to these hazards that originate from personal factors.

Performing an inventory of this depth does not have to be nor should it be a one-person endeavor. This is an initial activity that serves as a prelude to planning and making organizational change. Utilizing the personnel who have knowledge of the areas and activities involved in the inventory helps prevent overlooking important details and also gets personnel involved in the workplace safety endeavor.

REFLECTION 3.1

1. Mature safety management systems measure success by the hazards eliminated (Kausek 2007). Can you develop a means for tracking hazards by location or activity from the initial inventory process as a means to measure overall program success? What would the endeavor entail?

2. In previous chapters it was suggested that management philosophy that identified customers and satisfied their needs with services was a window to assessing the scope of the system for core management programs. In what degree does an inventory for hazards, emergency events, and vulnerabilities serve to develop the scope of the overall system?

ASSESSING FOR THREATS AND VULNERABILITIES

Recognizing vulnerabilities is the most basic of security skills. Vulnerabilities are conditions that can be exploited to breach security in order to gain access to an asset. Akin to a vulnerability is a threat. A threat is a potential action involving exposure to a hazard that now has human culpability or a level of intent. A good example of a vulnerability is an unlocked door. A good example of a threat is an active shooter. Anticipating reasonable threats and recognizing vulnerabilities and anticipating their exploitation are basic skills for tactical personnel and in managing security. It is important to realize that occurrences such as wildfire, tornadoes, or flooding are not threats, but rather natural disasters and are included as natural disasters when discussing emergencies in general, such as an electrical fire. They must be planned for as well, but are not the same as a threat. In the situation where a manager of workplace safety does not have stand-alone responsibility for security, he or she must be competent in the area of security to interface with security personnel, perform emergency response planning, and coordinate with company personnel and outside agencies in response to security and emergency events.

Security is truly the protection of any asset through five stages. These stages are deterrence, detection, delay, response, and rehabilitation (Philpott 2010). A vulnerability is a condition that allows these to be defeated. Critical to security is the concept of barriers. Barriers is a concept that I developed from the concepts of cover and concealment and from building defensive positions with the US Marine Corps in order to teach tactical positioning to police personnel when dealing with human subjects. But barriers are also critical in security. Each level of security has a goal, and barriers must be deployed at these points or levels congruent with the goal to be effective.

Physical barriers are one of three types: strategical, tactical, or protective. It is the goal of its use that determines its type. A physical barrier can be deployed in a strategic manner that means to deter movement or access by that route. A physical barrier can be deployed in a tactical manner that means to slow and is also used to aid in detection. Or, the same physical barrier can be used as a final protective means that is designed to be very difficult to penetrate and allow time for adequate response. A great example, although maybe extreme, are mines. A mine field can be lain that is very visible and deliberate to influence an enemy force to avoid movement via that route. But the mines may be lain in a manner that detects and slows advancement to maximize chosen defensive or offensive tactics. The example continues as using artillery to pound enemy forces as they advance against the tactically deployed mines. The same barrier can be deployed in close proximity to act as a final defense or protective barrier as well. The example continues with mines being deployed to protect a place of limited view near a soldier's fighting hole. In a less extreme example, a hedgerow of landscaping can be used to deter people from leaving a sidewalk prior to an entry door or in safety prior to a pedestrian crossing. This is an example of a strategical barrier. The same hedgerow may be deployed tactically to slow or delay persons trying to look into or climb through a window. Many times benches for sitting are arranged near entrances as protective barriers meant to stop vehicles from penetrating the entrance if an accident occurs or if a person intentionally wants to use a vehicle as a battering ram.

Deterrence is the outer-most level if viewed mentally as a perimeter, or the lowest level in terms of complexity to defeat or even be effective. Deterrence is a psychological approach or barrier for security. Deterrence can be obtained with psychological communication and with physical barriers that combine goals from successive levels of security. Cameras, for example, when deployed at the deterrence level are visible and signage warns of their use. Many fake cameras may also be used. A fence used as delaying barrier at the outer perimeter serves both purposes: to deter and to delay. Rather than skip detection level goals, the fence may also be used in conjunction with vibration or other sensors that now allow detection. But these examples cover typical deterrence level usage of barriers.

Detection is any measure that allows detection of entry. Barriers can be used with sensors and alarms to detect an approach, like the fence deployed simultaneously with the cameras at the deterrence level. But we also have detection points where we may use metal detectors, physical inspection, and bomb- or drug-sniffing dogs. For example, most schools today have a front entrance door that must be unlocked by a person after a visitor presses a button and announces his or her desire and identity to enter. The person is then directed immediately to a door that leads to the front office without any access to the school hallways.

Delaying level deployment measures include locked doors, fences, gates, and combination or keyed locks and are deployed after detection points and delayed or slow entry when entry is not desired so that time for tactical response is gained.

Response is the physical and/or personnel response aimed at stopping access to an asset. It may be that after detection an item to be protected is secured to a concrete floor and time expected to defeat the measures is greater than typical police response to stop the theft. Here, delaying measures and response work together. But response must be planned and calculated with consideration of the previous levels of security.

Rehabilitation can have measures that gather evidence for prosecution and gain information for assessment to improve security. Cameras deployed at convenient marts above the counter are deployed in the rehabilitation stage. The deterrence measure becomes the sticker on the entrance door that warns of camera use. However, the inadequacy of this system becomes evident when examined from the view of matching barriers and measures with the level and goal of security. The deterrence measure is deployed too late at the entry point without any measure for detection of an unwanted guest, nor delay once entry is made.

In conducting a security inventory, we utilize the concept of security levels and barriers to assess physical security. We must identify current points and measures and map them to coordinate security planning with emergency response planning. In the beginning we will utilize the perimeter model or view of security, and as we work closer to the asset of protection we will consider more of the effectiveness view of security. Table 3.15 suggests one such format for identifying barriers, measures, and points for coordinating the security effort. This will help us assess whether the barrier or measure is deployed consistently with the goal and is used in coordination with other barriers or measures.

TABLE 3.15 *Security Inventory Template*

SECURITY LEVEL	GOAL	POINT	BARRIER/ MEASURE	HOW DEPLOYED

Safety managers must assess their facility for vulnerabilities and threats in the manner of a hazard inventory. This begins at the outer edge of the property. From there, assess as to where entry is desired. Security is about providing access for desired personnel and preventing access to undesired personnel. Therefore, entry points are a first concern for identification. Assessing perimeter security first, the assessment works inward toward the center of assets of primary concern. Points where signage, cameras, dummy cameras, fencing, or any barrier is used, must be identified. Consider the following examples utilizing the security inventory format from Table 3.15 in Table 3.16.

TABLE 3.16 *Security Inventory*

SECURITY LEVEL	GOAL	POINT	BARRIER/MEASURE	HOW DEPLOYED
Deterrence	Deter entry to grounds	All perimeter	Fence with warning signage	At property edge with signage
Deterrence/ detection	Deter/detect	At east end of parking lot/ Lynde Ave.	Manned gate point	Manned gate point
Deterrence	Deter	Along front sidewalk to main office entrance	Shrubs	1. Discourage approach to building grounds/ windows 2. Guide toward main entrance

This example shows some possible entries and use of barriers identified as location points coordinated with security levels and goals matching their deployment. This serves as a master list to cross-reference numbered points on a property map. This will help in future planning for security needs. The assessment begins at the outer perimeter and progresses to entry points, and possible access areas to the asset. The investigator must be able to predict possible approach points and envision how to deploy barriers and measures to counter access.

NATURAL DISASTER INVENTORY

Natural occurring emergencies are planned for from a reasonable legal standard. The occurrence must be commonly planned for by industry in close proximity and be reasonable in terms of past or probable occurrence. A safety professional can examine records from the National Weather Service, National Oceanographic and Atmospheric Association, and from local regional dispatch centers for common and reasonable natural occurring emergencies. These emergencies can include earthquakes, flooding, wildfires, and tornadoes. But certain regions are more prone to different emergencies than others. For example, on the West Coast of United States, wildfire is a more frequent and reasonable threat than on the East Coast. However, the eastern coastal area is prone to hurricane events.

Table 3.17 shows one format for possible natural disaster events. The inventory is specific to the actual location of the facility or site to be protected; however, it is important to note that natural disaster impacts an organization through its workforce exposure as well. In determining risk to the actual facility, probability can be quantified through history of occurrence and severity by design or controls in preparation for the event. Some facilities are newer and built with earthquake resistance in mind, for example, or the event may not impact the facility with total shutdown or destruction. Not all of the facility grounds may be exposed to issues such as flooding, for example.

TABLE 3.17   *Natural Disaster Inventory*

EMERGENCY EVENT	SOURCE OF DATA	RISK
	Local agency: Y☐ N☐	Probability: L☐ M☐ H☐
	Agency:	Last known occurrence:
	Weather service: Y☐ N☐	Severity: L☐ M☐ H☐
	NOAA: Y☐ N☐	Exposure: L☐ M☐ H☐
	Area industry: Y☐ N☐ companies:	

THREAT INVENTORY

Threats involve human intent and varying levels of creativity, application, and commitment. Threats are more difficult to anticipate because of technology and cultural changes that affect the culpability and design of the threat. For example, since 1993, school shootings have become more common and are now a common threat prepared for by school districts in Kentucky. However, the first known school shooting in Kentucky occurred in 1902. In the late 90s and early 2000s, the threat of anthrax in mailings or suspicious packages was a common threat, with many actual acts or mistaken threat events occurring. In the early 1970s, civil protests expanded with several bombings that occurred and the level of concern or recognition for such threat increased accordingly.

Sometimes threats are masked by the nature of the event, such as arson. It may not be readily apparent at first that arson has occurred, but it would be wise to be aware of the possibility so that this is a consideration reflected in safety policy. An example is the banning of or even inspection of worker personnel for ignition source devices in certain areas where strict control of ignition sources are necessary. An example would include facilities or facility areas that are susceptible to build-up of combustible or flammable materials.

The whole idea is to be able to anticipate threats to develop an emergency response plan that either specifically predicts and prepares for the threat or is adaptable and accounts or helps in uncovering and preventing unpredictable behaviors and situations.

A good example of an adaptable security program is one that utilizes communication of threat level and increased security measures according to level. Green-level conditions may reflect no known credible threat, yellow may indicate a possible threat, orange may indicate a credible threat, and red may indicate an imminent threat. As the level increases, additional checks or tighter checks may be required. For example if the threat level was orange, area supervisors may be required to perform a door check and area check at an increased frequency. In a manufacturing plant setting, some bay doors are raised for ventilation. These types of doors usually have gates that can be locked as well. Therefore, a red threat may require the bay doors be closed to prevent a clear view into the plant. Programs like this are very valuable to the personnel who are involved in increasing security measures. Other personnel may not realize the value because they are not directly involved with increased responsibility. Competent security personnel may also be privileged to information that cannot be or should not be disseminated to all personnel.

Because security and emergency response planning involves sensitive information, competent-level personnel should also be from the management team or professional security services and contractors. This also preserves legal liability requirements centering on privacy. For example, as the threat level increases, personal information for the suspect or other sensitive information may need to be given to competent-level security personnel. In dealing with a domestic violence situation, the person who had committed a domestic assault had not been arrested by the time the victim had returned to work from her days off. The suspect was wanted and we disseminated his picture and vehicle picture and description to all supervisors and managers. The threat level was increased to orange. We also adapted a procedure for her pickup and drop off at work. Her parents were transporting her until the situation was over so we pulled the vehicle in one bay door and allowed it to exit another to minimize possible exposure of her and our personnel.

ASSESSING THREATS AND VULNERABILITIES

The successful safety professional must have a foundational grasp of recognizing hazards, threats, and vulnerabilities, as well as an understanding of abatement strategy. Assessing vulnerabilities and threats is not unlike hazard recognition. The process begins with identifying the asset to be protected and rating its criticality. It moves to identification of threats and vulnerabilities, risk rating of threats and vulnerabilities, and finally the identification and assessment of counters. At the foundation of this assessment is the ability to identify threats and vulnerabilities. They are very different because they are less predictable and they also rely on hazard recognition.

Recognizing vulnerability is best approached by recognizing the two definitions of vulnerability. The first definition includes conditions that can be exploited with culpability to gain access and defeat security. In other words, it is a vulnerability to a threat. There are four categories of vulnerability to a culpable threat: approach, detection, access, and movement.

Let's first examine approach vulnerability from the stand point of looking from point A for an advancing person. What allows approach? It is usually unfettered avenues. Our goal is to allow approach of welcomed non-threats while minimizing approach by a threat. This must involve establishing procedures that allow for determining an out-of-place action. Therefore, it involves the first level of security, or deterrence. At this level we would utilize barriers such as distance, fencing, or landscaping to funnel approach to a manageable point or points. Recognizing vulnerabilities of approach then would involve a lack of strategical controls influencing approach.

The strategical controls for approach vulnerabilities need to be examined for characteristics and the counters that can be deployed to meet or withstand the characteristics. Is the approach from passenger vehicle, human subject, commercial truck, or even a tactical vehicle? Counters deployed strategically do not necessarily have to withstand the threat but must encourage a path of travel and make detection of ill intent easier.

Detection at the earliest reasonable distance is desired. The point at which separation of local traffic and traffic that is relevant for accessing an area of protection must be considered. Will detection be necessary at the outer perimeter? Numerous devices exist to aid in detection: vibration detection, movement detection via camera monitoring, software for automated camera monitoring, alarms, and others. Early warning and redundant monitoring is critical. A common mistake is to have detection deployed at the access point alone. When possible, early warning to a controlled access point should be deployed. Let's look at a locked door that requires announcement by the person wanting access and then a deployed unlocking by monitoring personnel. It would be best to have early warning and visual on the person approaching. A pressure mat may give audible indication to view camera visuals in that area. Subjects with ill intent may defeat access control at that final point. Early detection may give advanced warning of intent in the case where visuals are deployed. A very common mistake is to deploy cameras at the detection stage but use them as deterrence or for capturing evidence for the rehabilitation stage, so it is vital to assess for redundant deployment of detection at an early distance.

Access vulnerability relies on the adequate deployment of barriers that impede threat progress to entry and progression once entry is gained. Progression after access is movement. A barrier is a concept of physical and nonphysical items that impede a threat to its target. It can be doors, locks, impact resistant glass, screens, bars, desks, fencing, or any physical item that is between the threat and the asset. But it can also be nonphysical and physiological. Distance, lighting, noise, and positioning are examples of nonphysical barriers. Assessing access and movement vulnerability is also dependent on barriers deployed at the access level and beyond. They must certainly impede a threat long enough to facilitate response.

Vulnerabilities exist as weakness to hazards. These are the physical conditions that allow an increased potential for hazard exposure. A basic example would be the environmental hazard presented by weather such as a tornado. Weak or poorly designed and constructed walls present an obvious vulnerability. Other examples might include non-safety glass used at potential entry points or near areas of close human contact. Here, we take the strategies that Dr. Haddon had for preventing injury and apply them to environmental energy exchange to humans and property. Basic hazard analysis extends to property as well as human subjects.

In assessing vulnerability to hazards we can examine the known or predictable characteristics of the hazard, such as temperature or force, and examine the barrier or other strategy that protects the asset for capability of withstanding the characteristics of the hazard. Vulnerabilities apply to any asset, structural or human. In abating a heat hazard, for example, we must understand the limits of human capability, or anthropometrics. We may be able to reduce temperature exposure from a heated manufactured item to an acceptable range for direct contact with the skin. Otherwise, we would have to match glove temperature protection ranges to the temperature range—the same as reinforcing door structures to withstand a certain level of force that might be deployed against it.

Vulnerabilities can also be simple weaknesses, such as an open window that provides easy access. Barriers can be deployed in similar situations that reduce the ease of access to potential weak points. Shrubbery that hinders easy access to windows might be a proactive approach. Couple shrubbery with height or distance from the ground and access by window is hindered and requires more complex strategies to defeat.

Identifying hazards and threats can begin with examining the historical occurrences at the facility, organization, and area. Threats must also be predicted based on current cultural trends and the examination of possible scenarios, to include those not reasonable. Vulnerability identification relies on hazard and threat identification. It is then that we can examine structural vulnerability from the aspect of approach, detection, access, and movement. Vulnerabilities can be identified and barriers examined for matching the characteristics presented for protection. Consider Table 3.18 for an inventory assessment of vulnerability.

TABLE 3.18 *Vulnerability Inventory*

CATEGORY OF VULNERABILITY	IDENTIFICATION CRITERIA	IDENTIFICATION OF COUNTERS	CHARACTERISTICS OF HAZARD/ THREAT	CHARACTERISTICS OF COUNTER	NEEDED CHARACTERISTIC OF COUNTER
Approach	Lack of strategical controls				
Detection	Early detection capability *Redundant detection counters				
Access	Barriers to access: Tactical and protective deployment				
Movement	Barriers to movement: Tactical and protective deployment				

Source: Dotson, Rawlins, Blair, and Rockwell 2017.

COUNTERING THREATS

Threats and vulnerabilities are fundamentally different from hazards. Threats and vulnerabilities involve human culpability. Threats are direct or indirect indications of harm from a level of human mental state. They are human directed. There are generally four levels of mental state: recklessness, wanton, knowing, and intentional. These states are motivational in human action to any incident. The states are not applied to the human action themselves, but to the results or outcomes and intended outcomes.

The state of recklessness exists when a person does not recognize that his or her actions will result in harm. Wanton activity exists when a person knows that his or her actions could result in harm but chooses to conduct the action anyway. Knowing activity is when a person knows that his or her actions will result in harm to another. Intentional is when the result or outcome was fully intended. The difference between wanton, knowing, and intentional may be better demonstrated by saying that in wanton, conduct harm was not intended. It was possible, but the person took a chance. The difference between knowing and intentional is that a person may have known or even wanted to create harm, but did not intend for the specific outcome that actually resulted from the action. Intentional is when a person means to commit and in fact committed a specific harm.

I have seen others attempt to apply culpability toward workplace incidents. I also believe in directly tying mental state to workplace incidents in regard to deciding the level of any disciplinary action. However, it must be applied to the result and it must be recognized that if the safety department has and is developing safety competency with effective training and experience, most unsafe acts will be wanton in nature. If it is beyond wanton, it becomes a criminal act. The application of mental state in workplace investigations is an advanced investigational skill. It is and should be the primary focus of any law enforcement agency responding to workplace incidents. This determines whether it will take over the investigation or whether it becomes a civil investigation for federal or state OSHA agencies and the employer.

Countering a threat is fundamentally different in overall philosophy than abatement of a hazard. Hazard abatement places an emphasis on avoidance of exposure. Countering a threat means that exposure is faced while mitigating harm. Let's compare that to hazard abatement schemes. In Table 3.19 you can see hazard abatement based on the ANSI hierarchy of controls compared to active countering of threats using the STP model for tactical personnel.

TABLE 3.19 *Hazard Abatement*

TYPICAL HIERARCHY OF CONTROLS	STRATEGICAL, TACTICAL, PROTECTIVE SAFETY
Elimination of hazards Substitution of dangers	Elimination of threatening person(s) Direct elimination/indirect elimination Incapacitation of threatening person(s) Direct incapacitation/indirect incapacitation
Engineering-level controls of hazards to eliminate or minimize exposure	Elimination of threatening person(s) Direct elimination/indirect elimination Incapacitation of threatening person(s) Direct incapacitation/indirect incapacitation
Administrative control of exposure	Administrative controls toward mitigating harm
Personal protective equipment or barriers to mitigate exposure	Protective gear and barriers to minimize harm to tactical personnel, separate from harm

In this table you can see that in dealing with threats, the probability of the threat continuing or exposing additional harm to more persons or property warrants or requires direct or indirect confrontation. The goal is to confront and stop the imminent threat of harm while also protecting responding personnel as much as possible. Countering threats means that effectiveness is also efficiency. Elimination or incapacitation of a threat involves either direct or indirect confrontation. A good example is a drone, or armed robot, as opposed to direct human contact. Administrative controls to a threat might involve isolation of the threat or area evacuation, but it might include administrative controls at the responding personnel in scheduling times on duty, observation time, superior numbers, and even training. Here, though, the harm is not avoided necessarily, but results of exposure are minimized. Protective gear level includes clothing, bulletproof vest and eye protection, and many common items to safety professionals, but might also include bulletproof shields or armor protection to counter harm from direct confrontation.

The level of harm being displayed as potential or probable determines the use of legal force for law enforcement personnel, and to a varying degree military personnel. It's worth covering to fully understand countering a threat. The use of force continuum is widely accepted and recognized in state laws and taught in training academies across the nation. It is used to justify the level of force one person uses on another. Due to training in the use of force and defensive tactics, law enforcement personnel are allowed to exhibit one level of force above the level of resistance displayed by the threatening subject, as long as they are acting within the color of law to arrest or legally detain a person.

The scale begins with psychological intimidation. Examples might be clothing, facial expressions, or markings. Equal to this is presence by a trained responder. The next level of resistance is verbal noncompliance and equal in the scale is verbal direction by the responder. For a trained responder, the next level, the plus-one application, is to utilize soft empty-hand holds. This level of force is a physical hold aimed at constraining, guiding, or directing movement. It is at this point that if a subject is progressive in his or her resistance to a responder things quickly escalate. This is because physical contact and confrontation must be administered. It is important to note that a person exhibiting a threat can enter the scale at any level. It is not always progressive. Most of the time it is not progressive. However, the longer the person is not taken into control, the more likely resistance is to progress. As resistance progresses with a threatening person, the likelihood of injury to a responder, others, and eventually the threatening person increases in direct proportionality to the level of resistance.

The scale progresses to passive resistance, and at this level the subject does not make verbal threats or perform any physical action other than limp body weight. The equal level of control is considered soft empty-hand controls or physical holds that restrain, constrain, or guide movement. For a trained responder, the next level, the plus-one application, is to utilize hard empty-hand techniques. This level of control is necessary to control the threat. Hard, empty hand might involve strikes, kicks to pain centers and balance centers, and decrease in vision to allow physical control to be obtained. An example might be a person running away and an officer tackling them. It could be a person pulling away and a swift kick technique to the leg, common peroneal area, used to take him or her down to put in handcuffs. The terms "soft" and "hard" denote the difference between striking to cause pain, distraction, loss of balance, vision, or will to resist and hold. Empty hand refers to the fact that the officer does not utilize impact weapons such as a baton.

Defensive resistance occurs when a person is jerking away, running away, or physically avoiding and dodging control but makes no attempt to strike or otherwise harm the responder. A good example is lying face down with hands clasping each other and preventing the hands from being freed for handcuffing or for confirming as empty of a weapon. Hard empty-hand control is considered an equal level of control response. The plus-one application for trained responders is to use intermediate weapons such as a baton to gain control.

The next level of resistance is active aggression. At this level of resistance the person is actively seeking to harm another. Hard empty-hand control is the equal control response. An example is a physical fight where strikes are being thrown. The plus-one application is an intermediate weapon such as a baton. Impact weapons are not an option for active aggression for untrained non-sworn personnel. These become clubs, which are considered a deadly force assault weapon.

Deadly force assault has an equal control of deadly force. Deadly force is not a plus-one application for active aggression, even for trained responders, unless extenuating circumstances exist. Weapons are not justified to be used by untrained, non-sworn personnel against unarmed subjects without meeting certain justifications. Since this is not about legal use of force, this text will stop with the basic explanation of the use of force continuum.

Countering a threat involves direct confrontation aimed at controlling the threatening person. Exposure is inevitable. It can be addressed with indirect exposure, but eventually physical possession or control is necessary. It reminds me of a common phrase with tactical personnel: "We run toward what you run from."

Vulnerabilities also involve culpability in that they are or can be exploited by persons to breach security. Vulnerabilities also include conditions that enable a threat or exist in weakness to another emergency, such as flooding or a tornado. But they most definitely are security related. Vulnerabilities might exist as unlocked doors, easily defeated security features, easy accessible points of entry, or even a lack of fire rating for walls.

Countering vulnerabilities involves matching the level of security and its goal to the deployment of the security feature. Just as in assessing for vulnerabilities, we divide the perimeter and approach the asset in the five levels of security: deterrence, detection, impedance, response, and rehabilitation (Philpott 2001). So, using the vulnerability assessment form found in Table 3.20, the developed counter must match the goal of its deployment in each of the categories of vulnerability.

TABLE 3.20 *Vulnerability Inventory*

CATEGORY OF VULNERABILITY	IDENTIFICATION CRITERIA	IDENTIFICATION OF COUNTERS	CHARACTERISTICS OF HAZARD/ THREAT	CHARACTERISTICS OF COUNTER	NEEDED CHARACTERISTIC OF COUNTER
Approach	Lack of strategical controls				
Detection	Early detection capability *Redundant detection counters				
Access	Barriers to access: Tactical and protective deployment				
Movement	Barriers to movement: Tactical and protective deployment				

Source: Dotson, Rawlins, Blair, and Rockwell 2017.

Therefore, we can list the characteristics of the counter as deterrence, detection, impedance, response, or rehabilitation, ensuring it is deployed accordingly. Let's take a common security control of camera use as an example. If we were to use a camera in the deterrence mode, we would not hide it. In fact, we would place signage advertising its use and possibly also deploy fake cameras as a means to deter. If we were using it as detection, the camera would have to be monitored by personnel and/or software technology that alerted us to movement or detection. Cameras may even be used to impede advancement. What if cameras were to be defeated by the assailants, cameras in number and in specific places could slow down advancement, giving more time for personnel to respond. Cameras are not appropriate for response, except to monitor events that are occurring. But cameras are an excellent tool for rehabilitation. Rehabilitation includes after-action reviews for improvement and also prosecution. So, cameras collecting evidence are primarily for rehabilitation. Of course, cameras can record evidence for use in prosecution, even when they are deployed as a deterrent. An example of improper camera deployment is when a hidden camera is used to record a customer and cashier exchange at a convenient store expecting the sign warning of camera use to deter the robbery.

As we explore the counters to increase security and overcome vulnerabilities, we must assess what the goal of the counter is and ensure its deployment at the right stage of security. Handling threats and vulnerabilities is performing safety. There is a real connection to safety and security. Unsecure places are not safe, and unsafe places are not secure. It may not be totally appropriate to run threat and vulnerability assessments similar to hazard recognition. Only personnel trusted with confidential information and of long-term stature can be involved with more advanced issues in regard to security; however, developing competency to the level of recognition and reporting of possible threats and of vulnerabilities is also very useful for the person responsible for this management area. Safety managers must at least work with peers managing these areas but are often tasked with managing these areas as well.

CONCLUSION

Safety professionals must be able to recognize the fundamental differences between hazards, threats, and vulnerabilities and inventory for the presence or possible presence of conditions or activities to understand the scope of the organization's needs. Abatement planning then begins with defining the specific requirements covered in the applicable regulations of jurisdiction. Compliance programs and policy can then be developed in the organization's preferred method. Abatement for hazards follows a general path of avoidance, whereas abatement of threats follows strategy, tactics, and personal protection gear concurrently, and, furthermore, abatement of vulnerabilities relies on combining the levels of security with a concept of barriers.

REFLECTION 3.2

1. Today's trend for threats includes active shooters in workplaces, schools, and areas of gathering. Examine this from the aspect of Bird's management model of causation and describe your primary abatement strategy.
2. Describe how you would involve workforce associates and management-level associates in performing vulnerability and threat assessments. If you prefer not to involve them, explain why.

REFERENCES

Dotson, Ron, Troy Rawlins, Earl Blair, and Scott Rockwell. 2017. *Principles of Occupational Safety Management.* San Diego, CA: Cognella.

Kausek, Joe. 2007. *OHSAS 18001: Designing and Implementing an Effective Health and Safety Management System.* Lanham, MD: Government Institutes.

Occupational Safety and Health Administration. 2018. *General Industry Regulations,* 29 CFR 1910. Washington DC: US Department of Labor.

Philpott, Don. 2010. *School Security.* Longboat Key, FL: Government Training Inc.

US Bureau of Labor Statistics. n.d. "Industry at a Glance." https://www.bls.gov/iag/home.htm

Figure Credits

Table 3.1: Adapted from OSHA, "Subpart D Inventory Strategy," Title 29 of the Code of Federal Regulations, Part 1910,.

Table 3.2: Adapted from OSHA, "Subpart D Inventory," Title 29 of the Code of Federal Regulations, Part 1910.

Table 3.3: Adapted from OSHA, "Subpart F Inventory," Title 29 of the Code of Federal Regulations, Part 1910.

Table 3.4: Adapted from OSHA, "Subpart H Inventory," Title 29 of the Code of Federal Regulations, Part 1910.

Table 3.5: Adapted from OSHA, "Subpart I Inventory," Title 29 of the Code of Federal Regulations, Part 1910.

Table 3.6: Adapted from OSHA, "Subpart J Inventory Topics," Title 29 of the Code of Federal Regulations, Part 1910.

Table 3.7: Adapted from OSHA, "Subpart J Inventory," Title 29 of the Code of Federal Regulations, Part 1910.

Table 3.8: Adapted from OSHA, "Subpart K Inventory," Title 29 of the Code of Federal Regulations, Part 1910.

Table 3.9: Adapted from OSHA, "Subpart L Inventory," Title 29 of the Code of Federal Regulations, Part 1910.

Table 3.10: Adapted from OSHA, "Subpart M Inventory," Title 29 of the Code of Federal Regulations, Part 1910.

Table 3.11: Adapted from OSHA, "Subpart N Inventory," Title 29 of the Code of Federal Regulations, Part 1910.

Table 3.12: Adapted from OSHA, "Subpart O Inventory," Title 29 of the Code of Federal Regulations, Part 1910.

Fig. 3.1: Source: https://commons.wikimedia.org/wiki/File:FCAW_drawing.JPG.

Fig. 3.2: Source: https://commons.wikimedia.org/wiki/File:SMAW_weld_area.svg.

Fig. 3.3: Source: https://commons.wikimedia.org/wiki/File:Puntlasmachine.jpg.

Table 3.13: Adapted from OSHA, "Subpart Q Inventory," Title 29 of the Code of Federal Regulations, Part 1910.

Table 3.14: Adapted from OSHA, "Subpart S Inventory," Title 29 of the Code of Federal Regulations, Part 1910.

Table 3.18: Ron Dotson, Troy Rawlins, Earl Blair, and Scott Rockwell, "Vulnerability Inventory," Principles of Occupational Safety Management, pp. 36. Copyright © 2017 by Cognella, Inc.

Table 3.20: Ron Dotson, Troy Rawlins, Earl Blair, and Scott Rockwell, "Vulnerability Inventory," Principles of Occupational Safety Management, pp. 36. Copyright © 2017 by Cognella, Inc.

CHAPTER FOUR

Constructing a Three-Tiered Hazard Recognition Program

FOREWORD

The hazard recognition program is where the "rubber meets the road" for the occupational safety program of an organization. In Employee Oriented Safety this means that hazard recognition does not merely include associate input; it also has components that rely on associate participation. This means that associate participation for performing the most basic safety duty is measured and helps define the organization's culture. The hazard recognition program is a three-tiered program: workforce associate recognition and reporting, competent safety associate recognition and abatement, and managerial-level competency. The program will rely on the fundamentals of recognizing hazards and then dividing hazards into levels that are appropriate to competency.

Hazards, threats, and vulnerabilities may be similar in that each increases the potential for negative harm. However, each is distinctly different in the culpability of harm and the abatement strategy. Hazards are physical conditions that increase the risk for injury or illness in humans. Environmental hazards also carry potential harm to plant, animal, and soil but are called "aspects." Threats involve mental culpability, or the state of the mind in regard to the results of action. In other words, it can be intentional. A specific example may include someone physically assaulting another. Vulnerabilities are conditions that exist that increase the potential for a breach of security. While this chapter will focus on hazard recognition, it is important to distinguish between the three, because safety professionals must recognize all three. Safety and security are not separate concepts at all. One cannot exist without the other. Recognition of the three is the most basic of investigational skill.

Objectives

After exploring this chapter the learner will be able to do the following:

1. Differentiate the differences between hazards, threat, and vulnerabilities
2. Apply the categories of hazard toward hazard recognition training
3. Distinguish the levels of hazard based on competency
4. Differentiate the goals of each tier of a hazard recognition program
5. Develop a three-tiered hazard recognition program

LEARNING PLAN

LEVEL	ULTIMATE OUTCOME	CLAIM	LEARNING TYPE	ASSESSMENT
F	Differentiate the differences between hazards, threats, and vulnerabilities	Hazards, threats, and vulnerabilities are three distinctly different challenges that safety managers must be able to recognize and counter.	AC	
U	Apply the categories of hazard toward hazard recognition training Distinguish the levels of hazard based on competency	Workplace safety management begins with understanding the legal liability of the organization in relation to its activity.	IL/AC	Learners will develop a general training content outline for a three-tiered hazard recognition program that corresponds to the levels of hazard and personnel competency.
U	Differentiate the goals of each tier of a hazard recognition program Develop a three-tiered hazard recognition program	Safety managers must development a hazard recognition program that develops competency and measures accountability.	CT	The learner will develop a core management program that reflects a three-tiered hazard recognition program.

Learning Plan Legend

Level:

F: <u>Foundational outcomes</u>: Basic abilities

M: <u>Mediating outcomes</u>: Progress through a developmental model; interpret, analyze, evaluate progressively challenging claims, arguments

U: <u>Ultimate outcome</u>: Navigate most advanced arguments and claims

Type of learning:

CR: <u>Critical reading</u>: The ability to read, process, and understand the meaning of written information

IL: <u>Information literacy</u>: Locating and selecting suitable information for a task; evaluating appropriateness/validity of information sources

AC: <u>Application of concepts</u>: Ability to apply discipline-specific knowledge/skill to tasks/situations important to the discipline

CT: <u>Critical thinking</u>: Ability to apply a concept to a vague or argumentative claim without a creative leap

AT: <u>Analytical thinking</u>: Ability to critique/analyze situations using a concept or model

CA: <u>Creative application</u>: Ability to apply a model/concept in a new way/to an unrelated situation or scenario; involves creative leaps

DEFINING HAZARDS, THREATS, AND VULNERABILITIES

In the examination of hazards and in developing the skill of workforce associates to recognize those hazards, it is clear we must break hazards down into a manageable number of categories that have little overlap. Each category of hazard represents the type category of hazard for which associates should be on alert. This is in contradiction to teaching two types of hazards or few options and expecting workforce associates to remember all the types and specific sources that fall into the limited options. The list that follows identifies the categories of hazards.

- Impact (hazards that hit or strike a person)
- Penetration (hazards that cut or puncture a person)
- Compression (hazards that crush or pinch)
- Chemical (hazards that can result in multiple types of injuries resulting from exposure to hazardous materials)
- Respiratory (hazards that interrupt the whole-body function of breathing)
- Temperature (conditions that result from exposure to hot surfaces or ambient air)
- Visibility (hazards that result from low visibility or contrast)

- Radiation (hazards that originate from exposure to radiation including the sun)
- Walking/working surface (hazards that result from substandard surface conditions)
- Electricity (hazards that result from exposure to electricity)
- Animal, insect, vermin (hazards resulting from exposure to animal, insect, or vermin)
- Biological (hazards resulting from exposure to bacteria or virus)
- Noise (hazards that result from noise exposure)
- Ergonomic (hazards resulting in musculoskeletal disorders)

These categories provide specific definitions that limit overlap and facilitate recognition of multiple sources. Ergonomic hazards could include all hazards as the field of human factors is principled with human-centered, designed workstations. But the meaning has been coded to cull out more specifically different conditions. Visibility, noise, and temperature are basic ergonomic concerns; however, we are designing ways for workforce associates with less competency in hazard recognition to find and report the hazards that coincide with their level of responsibility.

Categories of hazards include types of incidents or sources. It is vital when preparing this hazard recognition training that specific sources be shown to the workforce that are present in the relevant facility. The hazard inventory is the source for this information. Table 4.1 provides possible examples.

TABLE 4.1   *Example Hazard Inventory*

CATEGORY	TYPE	SOURCE(S)
Impact	Struck by object	Boom of backhoe, vehicle/traffic, train, etc.
	Bump/run into object	Overhanging structure, access to tight space
	Vibration	Jackhammer use
Penetration	Sharp edges	Edges of metal
	Injection	Hydraulic oils under pressure
	Impalement	Uncapped rebar
Compression	Pinch	Ingoing nip points
	Entrapment	Heavy object that pins down or restricts
Chemical	Burn	Acid
	Toxicity	Carcinogen
Respiratory	Dust	Asbestos fiber, silica dust
	Oxygen deficiency	Displaced oxygen levels in confined spaces

Temperature/ weather	Cold	Walk-in freezers, outdoor temperature
	Heat	Sun, ovens, chemical reactions/fire
	Natural disasters/storms	
Radiation		Sun, lighting sources, radioactive elements, X-rays
Visibility	Contrast	Low lighting
	Blocked view	Blind spots
	Distortion	Looking into or through substance
Walking/working surfaces	Slippery	Contaminants on dry surface
	Tripping possibility	Uneven walkway, cords
	Height	Different levels
Electricity	Shock	Amperage, lightning
	Burn	Heat from resistance
	Electrocution	Amperage, lightning
Animal/insect/ vermin	Penetration	Physical penetration of teeth, fangs, stinger
	Poison	Toxic exposure to poison
Biological	Bacterial	Staph infections
	Viral	Hepatitis B, C, HIV
Noise		Loud machinery
		Concussion/blast waves
Ergonomic	**Physical:**	Strains from heavy lifting/improper lifts
	Overexertion	Typing, repetitive reaching/scanning
	Repetitive stress/trauma	Jackhammer usage
	Vibration	Hand in high position on controls
	Environmental demands	Too many audible indicators
	Mental demands	Stress

Source: Dotson, Rawlins, Blair, and Rockwell 2017

Hazards have results to humans. They result in physical and/or health hazards that are chronic or acute in nature. For example, chemical exposure may result in both a physical burn and a chronic health hazard, as shown in Table 4.2.

TABLE 4.2   *Example Hazard Results*

CATEGORY	TYPE	RESULT
Penetration	Injection of hydraulic fluid from pressure	Fluid injection injury with acute physical loss of finger and hand control; chronic effect of tissue toxicity
Chemical	Skin exposure: Trichloroethylene	Dry, irritation of skin as acute results Parkinson's disease from chronic health exposure

Source: Dotson, Rawlins, Blair, and Rockwell 2017

Injuries occur because energy from the environment is exchanged with humans in two ways. The first way is when the whole-body function is interrupted (e.g., drowning). The second manner is when a local part of the body experiences an energy exchange that violates the level of threshold the body can withstand. The greater the energy, the more severe the injury (Bird, Germain, and Clark 2003). Therefore, limiting or preventing this energy exchange is the key to injury prevention. Haddon's strategies for injury prevention are as follows:

1. To prevent the creation of the hazard in the first place
2. To reduce the amount of the hazard brought into being
3. To prevent release of the hazard that already exists
4. To modify the rate or spatial distribution of release of the hazard from its source
5. To separate in time and space the hazard and that which is to be protected
6. To separate the hazard and that which is to be protected by interposition of a material barrier
7. To modify relevant basic qualities of the hazard
8. To make that which is to be protected more resistant to damage from the hazard
9. To begin to counter the damage already done by the environmental hazard
10. To stabilize, repair, and rehabilitate the object of the damage (Bird et al. 2003, xi–xii)

However, another abatement strategy also exists and was practice with limited success in the first half of the twentieth century: accident prevention. Accident prevention originated from the line of thought that accidents were preventable and that education and training were the key to accomplishing prevention. In practice, accident prevention is the psychological side of safety (Bird et al. 2003). It involves awareness efforts, from training as well as from proper design of workstation load; it involves policy and procedural development to include rules and courtesies; and it involves hiring practices such as matching mental characteristics to job requirements.

Since both strategies have been utilized together, beginning in the late 1960s, as shown in the combining of both strategies in highway safety, the workplace has improved significantly. It is clear that abatement of hazards requires concurrent application of both strategies. Typically hazard abatement is taught as prioritizing a correction by implementing engineering controls first, such as those detailed by Dr. Haddon first, as they are more reliable due to minimizing human decision. Administrative controls that are secondary are defined as awareness and training efforts, along with scheduling techniques that limit exposure. Last is personal protective equipment. This is the least reliable since it involves the most human decision and does little to eliminate or substitute exposure. But truly administrative controls are, in practice, implemented concurrently with engineering or injury-prevention controls. It is near impossible to eliminate or substitute a hazard without having an accompanying training and education program due to advancing the development of hazard recognition and reporting with the workforce. It is key to empowering associates at all levels in control of their own safety.

In contrast to hazard abatement strategy, threat and vulnerabilities are overcome from a tactical safety viewpoint. The biggest difference is that exposure cannot always be eliminated or substituted. In dealing with a threat, for example, it may be more efficient or safe to all in the long run to take on the threat head on. We limit exposure or increase survivability through strategy, tactics, and personal protective equipment. Vulnerabilities are countered by using strategical, tactical, and final protective barriers to deter, detect, and slow a security breach so that a proper response can develop. Management initiatives that center on identifying and countering threats and vulnerabilities are as vital as hazard recognition and may be launched in a similar manner, as this chapter explores with hazard recognition.

REFLECTION 4.1

1. How does creating several categories of hazard help workers identify hazards more readily?
2. How do Haddon's strategies compare to the ANSI hazard abatement scale?

LEVELS OF HAZARD AND COMPETENCY

Every management program has three levels of competency: authorized, competent, and administrative. As skill is developed, responsibility increases. Identifying hazards in a workplace involves competency. There are certain hazards that the least-developed associate can recognize and correct and then there are those who require higher levels of competency. Level 1 hazards are those that all associates in an organization can recognize and correct, or at least know to report for corrective action. Then there are hazards that involve more advanced training and education. There are three levels of hazards all together. Level 2 hazards involve those that competent safety associates can recognize and plan correction for. Level 3 hazards involve advanced technical skill to identify or confirm and plan correction for. A good example centers on machine guarding. Any associate should recognize a loose barrier guard. A safety professional has the knowledge to find more advanced guarding requirements such as proper safety distance. But it might take a person with computer programming knowledge to test a production machine's programmable logic controller for faults, or a control engineer to properly program and install redundant controls for safety. Identifying all three levels of possible compliance issues or hazards is critical for a complete assessment or audit. These levels must be represented in any complete auditing initiative.

In other words, the audit program, or truly proactive investigational efforts, have audit instruments that identify all three levels of hazard. It will be critical to work with engineering and maintenance to identify the level 3 hazards or criteria in the audit instrument. Levels are important because if all associates are responsible for safety and empowerment is desired, the competency level of the associate must be measurable to the level of hazard. In other words, authorized safety associates, or the base level, should be competent and responsible for level 1 hazards. Competent-level safety associates must be responsible for planning and implementing countermeasures, as well as identifying level hazards. Special technical skill personnel must be responsible for level 3 hazards and conditions within their respective areas of expertise. This is the basis for safety professional development for organizational personnel.

Authorized personnel must have basic hazard recognition training and be allocated the duties to identify level 1 hazards, report the hazards, serve on countermeasure committees, and abide by all work procedures and safety policies. This means that hazard-recognition training should concentrate on the categories of hazard, specific sources in the facility or site, abatement of level 1 hazards, and how to report them.

Competent safety associates take on all duties and competencies of the authorized level but also must have training on abatement strategy, committee involvement and management, and enforcement strategy. It is possible and practical to develop workforce associates into competent-level safety associates when they are high-level safety performers, have safety as a virtue, and expected turnover is low.

Each program has an administrator level. This level has mastered the other levels but now is responsible for overall program success. This means that the administrator must be capable of identifying and tracking relevant metrics for constant adjustment and progress (CAP). These metrics will be used to report to peer and upper-level management levels on the success of the initiative and overall program. Table 4.3 suggests the matrix for the hazard recognition program.

TABLE 4.3 *Hazard Recognition Program Matrix*

COMPETENCY LEVEL	DUTIES	COMPETENCY
Authorized • Authority to work safe • Find/report hazards • Watch others	Recognize hazards Report hazards Report safety concerns Follow safety procedures Serve on select committees	Hazard recognition Hazard reporting process Level 1 abatement

Competent
• Authority to stop all unsafe activity
	Authorized responsibilities and the following:
	    Assess/plan countermeasures
	    Enforce procedures/policy
		Hazard abatement
		Problem-solving steps
		Committee management
		Experience
		Authority to stop work

Administrator
• Responsible for overall program performance
	Previous responsibilities and the following:
	    Review program performance
	    Formulates Continual Improvement Plans
	    Reports program status
		Performance Metrics
		Initiative Success
		Overall authority

Source: Dotson, Rawlins, Blair, and Rockwell, 2017

REFLECTION 4.2

1. How do you propose to track experience to identify competent-level personnel?
2. What is the problem with overlooking experience as a qualifier for competent personnel?

THREE TIERS OF HAZARD RECOGNITION

Employee Engagement

The hazard recognition program should have three main initiatives that approach this core management program. Tier 1 is an initiative that covers the immediate general safety conditions that are the most frequented level 1 hazards. Tier 2 is a job hazard analysis initiative. Tier 3 is a hazard recognition initiative that relies on employee participation. All three initiatives combine to comprehensively perform the most basic of safety functions. This is a three-tiered program because it engages workforce associates at the three levels required before competency allows for empowerment.

Workforce engagement has four levels: directive, consultative, participative, and empowered. Engagement begins at the directive level, which is where orders are very sequential in nature and are closely monitored for success with short-term goals of performance. The worker at this level is learning a new skillset and is not confident or motivated about the task. The first tier of the hazard recognition program covers the directive engagement level.

Consultative level engagement is input prior to decision making. At this level the worker is gaining experience with the new skillset and is motivated but not confident. The second tier of the hazard recognition utilizes input from the workforce associate in analyzing workstations for hazards.

Participation involves the workforce at a level where decisions to recognize, correct, and report hazards consistent with workforce-competency level are willingly exercised. A duty commensurate with competency is assigned and authority to act is granted. This describes the third tier of the hazard recognition program.

Empowerment is only possible when competency is fully developed. A successful three-tiered hazard recognition program develops workforce competency level to the point where actual decision making for safety at the personal and organizational levels can be effective.

Tier I

The tier 1 initiative is inspired from the old saying "First impressions are everything." My attitude in regard to inspections, developed in the Corps and in search protocol in policing, is that if the obvious or minor details cannot be mastered, dig deeper and you will find more complex and root issues. The purpose of this initiative is to eliminate the most frequently observed violations or hazards present, instill safety as a basic everyday responsibility and concern, disseminate knowledge from management to workforce associates, and provide an objective base for supervisor safety performance evaluations. It is the first initiative to be implemented.

The safety manager or team of competent-level safety personnel will perform walk-through assessments and, if possible during hazard inventory, with the purpose of identifying the first impression level 1 violations and hazards present and prevalent.

The safety manager or team should take the attitude of finding out what would make the first impression of a safety inspector a good one. The findings would be the ten or so most common violations or hazards present. Keep in mind that these may change over time. Examples include items leaned up so that overturn is probable, spills or slippery conditions, trip hazards, build-up of combustibles like cardboard, blocked exits, blocked electrical panels, mushroomed hammers, drift pins or other tools, housekeeping conditions, improper or no personal protective equipment, and missing fire extinguishers.

These basic level 1 hazards will be the foundation for developing hazard recognition skills among the workforce associates and the frontline supervisory personnel. It will also help establish safety as a virtue by becoming an objective measure for annual safety performance evaluation for supervisory personnel and workforce associates. The program will begin with supervisor performing one formal audit for these conditions weekly and submitting the results to safety. The results should in most cases reflect that correction was completed. However, some findings may require safety involvement or committee effort. These conditions are the responsibility of the supervisor on a daily basis. If safety identified the existence of one or more of these basic conditions on any given day, it becomes an instance of negative performance and should be documented. This can be tracked as a performance indicator and used to evaluate supervisors annually. After the supervisor has demonstrated consistent high performance, or lack of these conditions being present, the supervisor can delegate the weekly general safety audit to an authorized safety associate. The audit will still be submitted to the safety department to track associate participation and results. It will be encouraged and taught to the supervisor that he or she should have an authorized safety associate assist him or her during his or her tenure of conducting the audit. This mentoring strategy can also be used with the authorized associates. Associate performance can also be an element of report to all associates in regard to annual safety performance.

Tier II

The second tier of the hazard recognition program will consist of assessing each job and workstation for hazards. Job hazard analysis (JHA) will be conducted in conjunction with job task surveys (JTS). It will be the overall responsibility of the safety department to produce the official assessments and to post the JHA at each workstation's administrative area. Assessment will be completed by forming ad hoc committees and including workforce associate input.

The job task survey, or JTS, is a vital tool for interfacing duties of human resources and of safety. For example, a JTS can provide human resources and accounting with information such as what type of pre-employment physical is needed or what supplemental pays are to be awarded to safety shoe or other personal protective equipment supplements such as uniform allowances. But it is also especially useful in managing injured workers and providing official job assessments to the carrier or medical partner. A JTS will be completed for every official job title in the organization. It will be important for timely completion that human resources help in forming several committees that will include members of the actual position covered as well as members that perform similar duties. For example, when addressing front-office positions or departmental manager positions, a sitting member should be from that exact position, even if there is only one, and include members who are also in the front office and interface with that position. Departmental managers and a human resource manager can handle the committee for those positions. Even these positions should have a JTS on file. One big topic that affects worker's compensation premium is the time the manager is present in the production environment. This item is necessary to determine risk for work-related injury.

The JTS should include the minimum information displayed in the following box:

JOB TASK SURVEY

Position: _____ Date of survey: ___/___/_____

Company/Facility location: _____

JTS committee: _____

TABLE 4.4   *Joint Task Survey Information*

DESCRIPTION	TASKS	REQUIRED PPE	PRE-EMPLOYMENT	TRAINING REQUIREMENTS
General:		Pay supplements: ☐	General: ☐	
Time in production		Shoes/boots: ☐	Ergonomic baseline: ☐	
environment:		Uniform: ☐	Hearing: ☐	
Lifting:		Special: ☐	Vision: ☐	
Bending:			Color perception: ☐	
Crawling:			Depth perception: ☐	
Standing:			Background check: ☐	
Sitting:				

Each company can work with its worker's compensation carrier to identify the physical job descriptions elements that will be queried by the carrier for determination of off-duty cases. These descriptors of the position can help in determining the elements to be considered in the job description. Based on the position and the work environment, the company will have different pre-employment physical requirements. Medical partnership is a good practice for safety management. It is possible to have standard physical requirements set up to handle the different requirements. While some companies may give a comprehensive physical to all associates, others may prefer to conduct only the type of physical exam that meets the requirements of the work environment. Turnover rates can offset any advantage gained by conducting the same physical exam for all employees. For example, employees not exposed to the action level of noise per OSHA regulations are not required to have a baseline hearing exam. If the employee is retained long enough to move to a high-noise work environment, then a baseline hearing exam can be conducted. It is also important to note that a JHA may be required before all standard personal protective equipment for the position can be listed.

For each primary task that the position performs, a JHA will be produced. Table 4.5 shows possible JHA categories. In this system, desired and undesired behaviors will be included. Behaviors can change the dynamics of any hazard, threat, or vulnerability.

TABLE 4.5 *JHA Category Examples*

TASK	CRITICAL STEPS	HAZARDS	BEHAVIORS		COUNTERS
			DESIRED	UNDESIRED	

The basic JHA structure as pictured also serves as an observation instrument, especially for tactical personnel or for lab safety observers. It is a best practice for one person, usually the person assigned safety responsibility, to observe response at a scene such as a fire incident, police special response, or even during chemistry experiments in a school lab. Critical steps includes occurrences. When an occurrence is less than ideal, non-typical hazards can pop up. Does the response team follow proper protocol when the unexpected occurs? Do members of the group act in undesired ways? An after action review is truly practicing continuous improvement. On-scene observations identify weaknesses and produce an after-action review, which in turn identifies training and policy needs that can be improved. A possible example is when a police entry team confronts a threat and has one of its own becomes a casualty. Identifying what the team did next and how well it adjusted to the situation and correcting what actions were not ideal according to their strategy and tactics can improve training and preparation for the next similar occurrence. It is very similar in a manufacturing or construction organization when incorporating behavior-based safety, except that day-to-day work station observations should not have pop-up hazards as often or as severe. An example of a pop-up hazard in a manufacturing setting may involve a machine break down. Assessing whether the operator properly shutdown the machine and practiced his or her duties under the lock-out tag-out program is vital. A basic operator may in most cases practice stop, call, and wait—a process of normal shutdown, summoning a supervisor, and waiting to maintenance and further scheduling.

Including behaviors is vital for another reason in conducting behavior observations: It allows the observer to prepare for observing that specific workstation. Some workstations may have unique behaviors that can or should be targeted for review. Not all behaviors fit into standard categories such as failure to wear personal protective equipment. A workstation may have typical mistakes or shortcuts. It is the goal of behavior-based safety to more closely align designed protocols to actual occurrences. This involves moving both design and applications closer together. Observation of behavior is a method of hazard recognition and abatement itself. It is a component of job hazard analysis. Therefore, job hazard analysis and behavior observations combine for the second tier of hazard recognition. Behavior-based safety will be explored in another chapter.

Human factor programs, covered in a later chapter, can explore job hazards in another useful manner. Job tasks can be analyzed for demands. Physical and mental demands can be assessed as low, medium, or high once environmental conditions are identified. This adds even more depth to a job hazard analysis.

Tier III

The third tier of hazard recognition is a program that incorporates all workforce associates into the frontlines of hazard identification. It makes employee participation vital to the success of the third tier and also instills safety as a virtue, a true component of the equation that quality + safety= production. This level of the hazard-recognition program will be a primary point of tracking accurate participation by authorized-level associates.

The program will consist of authorized-level associates identifying and reporting level 1 hazards to competent-level associates. A hazard report card will field the hazard and the status of correction, from immediately corrected to referring to competent-level associates for the planning process. Depending on the maturity of the program or of organization structure and the preference of safety management, competent-level associates may be supervisory-level personnel or may include senior workforce associates who have demonstrated authorized-level performance and have the competency to fulfill competent-level duties. This could even be safety professionals assigned to work specific manufacturing lines or areas of the facility or site.

Once reported, the competent-level associate will assess the hazard and confirm it was reported accurately and confirm its level as 1, 2, or 3. When safety or other competent-level associates perform either compliance-level audits, program effectiveness audits, or merely walk through the area and discover a hazard, it will be queried for a previous hazard report card. If it is a level 1 hazard, it should have been recognized by an authorized-level associate rather than a competent-level associate, unless the hazard had appeared soon enough for the authorized associates to not have had adequate opportunity to discover it. Therefore, the program establishes credibility in its metrics for accurate participation. In other words, competent-level associates confirm that the authorized level is identifying and reporting or correcting the hazards that they are competent for identifying and correcting. We also establish that a lack of identifying and correcting or reporting level 1 hazards is a measure of lack of participation.

It serves as a program that requires safety participation. It is only one measure that will be used to calculate overall safety participation, but it is a primary measure of performance from one department or cross-functional grouping, such as areas or production lines, as compared to others. It can also serve to provide performance feedback individually and collectively for all associates. It becomes an objective measure to include in supervisory personnel evaluations, and workforce level evaluations. The measure could be done in multiple ways, however, I prefer to calculate the total level 1 hazards discovered by all levels of competency and use it as mark to measure the percentage of accurately identified and reported or corrected hazards by authorized associates. So, hazard recognition accuracy would equal the percentage of total level 1 hazards of a line or grouping that was correctly identified and corrected or reported by authorized personnel. This becomes a level 2 metric or operational leading indicator.

We have already established what culture is and what measures will be used to define and measure it. In Table 4.6, we have a charted list of measures for the cultural subset of safety. Notice that accuracy of hazard recognition is a measure of success for safety, demonstrating workforce commitment.

TABLE 4.6 *Safety Measures List*

LEVEL	ELEMENT	LEVEL
MANAGEMENT MEASURES		WORKFORCE MEASURES
Percentage of corrected safety problems from closed audits and project proposals	Commitment	Percentage of level 1 hazards self-identified and corrected.
Management representation and sign off on safety committees	Participation	Percentage of level 1 hazards identified by workforce level associates
Mean rating of perceptions from survey	Perception	Mean rating of perceptions from survey
Rating from ethical assessment of policy adherence	Compliance	Ratings from behavior-based observations as compared with authorized- and competent-level scores
Percentage of management developed to safety competent level	Competency	Percentage of workforce developed to competent level

There could also be other measures of workforce commitment, but once commitment is high and accuracy of hazard recognition is also high, the program is ready to be used as a basis for incentive. Commitment must be high or incentives can become self-defeating in that they may become viewed as a means for gain and not taken seriously.

Speaking in research terms, the program is conceptualized to answer the question, "Does a higher level of workforce participation in safety improve safety performance?" The hypothesis is that workforce participation in safety will result in a reduction of work-related injuries, workplace incidents, and costs associated with work-related injuries. We would hope to plot participation in the hazard-recognition program against these and other typical level 3 or lagging performance metrics to measure the effect and adjust management efforts to have the greatest positive result. This is the goal of management in general: to efficiently allocate resources to initiatives having the most positive effect on success. For this program, we hope to see a negatively correlated graph. As participation increases, injuries should fall. Figure 4.1 represents the anticipated positive outcome.

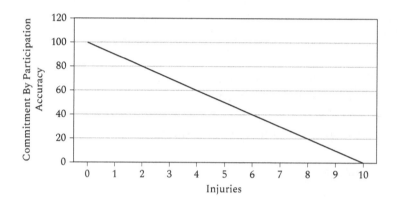

FIGURE 4.1   Participation 2

Of course, there are many variables that can impact injury rates. Employee participation and commitment are but two. It is also important to note that to make any causal connection between a newly implemented initiative and positive performance on final performance metrics, the safety manager must identify all the variables that impact the final performance measure. In the management of occupational safety, we know or assume that participation rates lead to better safety performance, and research backs this conclusion. But is it impacting a particular line, process, or workstation positively in this manner? To answer this type of question we must implement new initiatives in stages that can be easily divided for measure and conceptual practice. From the implementation of this third tier of the hazard recognition program, we have a control that increases the validity of data. The control is the inclusion of competent-level inspections and confirmation of the correct level assessment and correction.

There are more specific level 3 measures of performance for the hazard-recognition program that can be tracked. From the tier 1 portion we can track number of daily discrepancies and correlate it to participation. Here, the hypothesis is that as employee participation increases, basic discrepancies should fall, for example. The assessment of particular statistics from the hazard report card make possible the ability to gather data from the program. All programs must have tracked measures. Collectively these measures are the variables that impact performance metrics. For example, we might track investigational accuracy and countermeasure effectiveness from our investigations program as additional variables correlating to final performance metrics. Overtime changes in program measures and performance may reveal a level of correlation to performance.

Note the section of the hazard report card that is for safety department assessment. The box that follows displays the hazard report card.

HAZARD REPORT CARD

Location/Production Line:_____ Authorized Associate:_____

Date: __/__/____ Competent Safety Associate:_____

Hazard Category		Source	Control	Status
☐ Compression	☐ Radiation			☐ Corrected
☐ Penetration	☐ Walking/Working Surface			☐ Action Now
☐ Impact	☐ Electricity			
☐ Chemical	☐ Animal, Insect, Vermin			
☐ Respiratory	☐ Biological			
☐ Temperature	☐ Noise			
☐ Visibility	☐ Physical Ergonomic			
☐ Biological	☐ Physiological Ergonomic			

SAFETY DEPARTMENT/Competent Safety Associate USE ONLY

Hazard Found By Authorized Associate   IC #_____ Follow-up Date:__/__/____
Yes ☐   No ☐
Hazard Timely Reported   Hazard Level ☐1 ☐2 ☐3 Accurate Participation: Y☐ N☐
Yes ☐   No ☐
Problem Solving Committee Assigned
Yes ☐   No ☐
Countermeasure Plan Completed
Yes ☐   No ☐

Source: Dotson, Rawlins, Blair, and Rockwell 2017.

Each submitted card must be assessed by competent safety personnel. Then the data must be collected on a log that provides tracking of the results. Since the goal is to empower workforce associates to identify and correct or report hazards, targeting those hazards that match their level of competency, competent personnel must verify the hazard was identified by the authorized associate and confirm it was reported timely and coincides with competence level. If it was a hazard that was not immediately corrected, it must be action dated for follow-up on planning or results of the implementation of a countermeasure.

> **REFLECTION 4.3**
>
> 1. How does each tier of hazard recognition develop personnel competency toward recognizing hazards?
> 2. How does incremental implementation of the three tiers develop each cultural criteria?

PROGRAM METRICS

The first tier of the program is one possible measure for evaluation of supervisor performance. Since the basic list of first impression hazards or violations will be the daily responsibility of the supervisor to ensure the conditions are absent from the work environment, when such a condition is found it should be documented for purposes of objective safety evaluation. Safe conditions of the work environment is a valid responsibility of immediate supervision. Tracking the performance of supervisors to keep their assigned jurisdictions clear of these conditions provides an objective measure in safety evaluation.

In the third tier of the program, we are tracking employee participation but at the level of their competency. We can compare the number of level 1 hazards that have been found and reported by workforce-level associates to the overall number of level 1 hazards that are known. Competent-level personnel can and will audit for all levels of hazard, and when a level 1 hazard is found that has not been reported by the workforce level associates and that should have been found by the workforce-level employees, then it is an indication of a lack of participation. We can, over a period of time, compare the findings. Perhaps the workforce has only accurately recognized 75% of the known level 1 hazards found in their jurisdiction. We could make this the measure of participation for that jurisdiction, or at least one measure to be considered for participation. This means that audits and other types of inspections must be examined by the administrator of the program, for indications of level 1 hazards not recognized by the workforce. Table 4.7 suggest the formatting for a log of the third tier of the hazard-recognition program.

TABLE 4.7   *Hazard-Recognition Program Example*

JURISDICTION	IC#/HAZARD CATEGORY/SOURCE	RECOGNIZED ACCURATELY BY WORKFORCE
		Y☐ N☐

In this database for the third tier of hazard recognition, we could quickly calculate a percentage of level 1 hazards accurately recognized and reported by the workforce. This is a very important measure of culture through the behavior of hazard recognition.

Program effectiveness audits are a continual improvement effort that allows management to fulfill its goal of allocating resources at the right issue. Every program effectiveness audit has measurement criteria that follows the structure of management duties, operational and culture measures, and status or performance indicators.

Management duty criteria from all tiers of this program begin with allocating the necessary resources. This means the training and time to conduct necessary tasks. Management also has the duty to collect data report or give feedback to organizational members on their progress and performance. From a leadership perspective, aspect management must develop competency within each program to facilitate personal development and gratification of the employees.

Measures for the allocation of resources could center on training delivery percentages. Publication of annual or periodic reports on safety success to the organization's members is another possible measure for providing feedback. The second tier of the program involves job hazard analysis, and the percentage of workstations with up-to-date and reviewed JHAs is also a measure of management fulfilling its duties. It is possible to measure the increase or maintenance of competency levels in hazard recognition as well. If the level of competent personnel is growing or is maintained at an adequate number, this could be scored and indicate successful or inadequate accomplishment of this particular duty.

Operational and cultural criteria are measures of compliance to procedures, behaviors, and perceptions or attitudes. Surveying for the perceptions of this program is a good way to gauge its perceived importance. Surveys have their issues due to assuming honest responses. Safety can spot-check for occurrence of and adherence to procedure, such as having a method to check for each jurisdiction performing daily audits for the first impression hazards in the first tier of the program. Perhaps a procedure for keeping the latest audit is developed so that the administrator of the program can spot-check jurisdictions for proof of the daily audits.

Periodic Effectiveness Review
Management Criteria

ITEM #	CRITERIA	SCORE (1-5)
1	Percentage of Workforce level associates trained in hazard recognition	
2	Publication of safety reports to members of the company meets required schedule	
3	Percentage of workstations with up to date JHA reports posted	
4	Level of competent personnel is adequate for review of hazards/countermeasure review/level 1 audits	

Operational Criteria

ITEM #	CRITERIA	SCORE (1-5)
5	Mean score for supervisors perception of program importance	
6	Mean score for workforce associates perception of program importance	
7	Mean score for supervisors rating of company support for the hazard recognition program	
8	Mean score for the workforce associates' rating of company support for the hazard recognition program	
9	Adherence to daily conditions check	

Status Criteria

ITEM #	CRITERIA	SCORE (1-5)
10	Mean score of supervisor ratings for safe conditions of work environment (daily conditions check) Mean Rating: _____	
11	Mean score of supervisor ratings for safe conditions of work environment (daily checks) by jurisdiction: Xxx department: Yyy department: Zzz department:	
12	Employee participation rate for hazard recognition: Percentage: _____	
13	Employee participation rate for hazard recognition by jurisdiction Xxx process: Yyy production line: Zzz department:	

FIGURE 4.2 Periodic Effectiveness Review

Status criteria are the final performance measures for the program. It is important to realize that some of these metrics may be considered leading metrics when examined for overall safety performance of the organization. Status criteria for the hazard-recognition program begin with supervisor ratings on safe conditions of their work environment. We can average the score of supervisors' evaluation ratings, and it may be beneficial to report overall ratings of all supervisors and divide them into departments or other types of jurisdiction depending on organizational structure. Another important status indicator of this program comes from the review of hazard report cards. We can average the participation rate of all employees and of employees in specific jurisdictions.

Figure 4.2 is a basic template for program effectiveness review of the hazard-recognition program.

Of course, the parameters of scoring the criteria must be formally established in the procedures for conducting the program effectiveness review. This could change depending on the organization's outlook. But maybe we rate supervisors on the daily conditions check on a scale of 1 to 10. Maybe a finding of one discrepancy per quarter is worthy of a 9 score as an example of how to calculate their annual evaluation score. For item 10 we would calculate an average score of all supervisors. We could establish the scoring grade of 5 as being an average of 8.5 for the average of all supervisors' annual evaluation rating for safe working conditions.

REFLECTION 4.4

1. What is the problem with only assessing a program's effectiveness from status indicators?

CONCLUSION

The hazard-recognition program is the fundamental program for building a positive safety culture at the workforce associate level. It builds safety competency at the most basic level. Recognizing hazards is where the road to empowerment begins in safety. Workforce associates who cannot or do not recognize hazards, correct the most basic hazards, and report the conditions are not participative in safety. There can be several reasons why this occurs, but without the most basic safety skill, empowerment, or an active role in decision making, cannot be effective. Investigations skills begin with hazard recognition. Workforce associates are involved not only at the recognition stages, but in many safety-related duties that are investigational in nature, such as problem solving, observations, procedural development, and strategic goal setting, to name a few. The most basic investigational skill therefore cannot be lacking.

REFERENCES

Bird Frank E., Jr., George L. Germain, and Douglas M. Clark. 2003. *Practical Loss Control Leadership,* 3rd ed. Duluth, GA: Det Norske Veritas.

Dotson, Ron, Troy Rawlins, Earl Blair, and Scott Rockwell. 2017. *Principles of Occupational Safety Management.* San Diego, CA: Cognella.

Figure Credits

CHAPTER FIVE

Structuring Emergency Planning, Prevention, and Response

FOREWORD

After completing the initial inventory for hazards, security inventory, and vulnerability assessment, the next logical step is to begin planning for emergencies. This is the most important policy that will be developed and is one of the more complex plans to implement and ensure it is practiced as it is designed. Protection of this plan is also vital in that we should not communicate all details of this plan to the general public or to all associates.

Emergency planning and prevention planning is a combination of matching hazards, threats, and vulnerabilities to a plan for countering their development in the event that an emergency occurs. It combines communication and coordination of multiple activities and personnel assignments in a manner that preserves human resources first and limits loss of life, health and property. It is an effort that must take into account the capabilities of the organization and the capabilities, strategies, and tactics deployed by community-first responders to efficiently accomplish its goal. Simply stated, we predict, prevent, plan, and prepare for emergencies.

This chapter will cover the basic considerations and activities required to formulate such planning while introducing some of the basic strategies and tactics utilized by first responders.

Objectives
By the end of this chapter learners will be able to do the following:

- Identify emergencies required for response planning
- Develop a threat-level communication plan
- Understand common planning concerns surrounding threats
- Construct an emergency action plan
- Construct a fire-prevention plan
- Formulate rescue duties for unintended consequences
- Identify regulatory standards requiring specific emergency planning elements

LEARNING PLAN

LEVEL	ULTIMATE OUTCOME	CLAIM	LEARNING TYPE	ASSESSMENT
F	Identify emergencies required for response planning	Safety managers must prepare for emergencies associated with the location of their facility based on an assessment of probability.	AC	The learner will identify various emergencies for a fictitious facility based on the geological location where the learner currently resides.
U	Develop a threat level communication plan Understand common planning concerns surrounding threats	Threat-level communications can play a pivotal role in adjusting organizational security duties.	AC/CT	Learners will develop a threat-level communication plan corresponding to an increase in duty for critical personnel.
U	Construct an emergency action plan	Emergency action plans are a crucial element to organizational preparation for sustained safety.	CT	The learner will develop an emergency action plan that correlates to suggested practice, OSHA required elements, and EPA-required planning elements.
U	Construct a fire-prevention plan	Fire prevention planning is a critical element to organizational preparation for sustained safety.	CT	The learner will develop a fire-prevention plan.
U	Formulate rescue duties for unintended consequences	Formulating rescue duties for the type of activity and machinery is a critical preparation technique for safety management.	AC/CT	The learner will develop a policy for utilizing a committee to anticipate rescue scenarios based on activity and types of machinery utilized.

Learning Plan Legend

Level:

F: <u>Foundational outcomes</u>: Basic abilities

M: <u>Mediating outcomes</u>: Progress through a developmental model; interpret, analyze, evaluate progressively challenging claims and arguments

U: <u>Ultimate outcome</u>: Navigate most advanced arguments/claims

Type of learning:

CR: <u>Critical reading</u>: The ability to read, process, and understand the meaning of written information

IL: <u>Information literacy</u>: Locating and selecting suitable information for a task; evaluating appropriateness/validity of information sources

AC: <u>Application of concepts</u>: Ability to apply discipline-specific knowledge/skill to tasks/situations important to the discipline

CT: <u>Critical thinking</u>: Ability to apply a concept to a vague or argumentative claim without a creative leap

AT: <u>Analytical thinking</u>: Ability to critique/analyze situations using a concept or model

CA: <u>Creative application</u>: Ability to apply a model/concept in a new way/to an unrelated situation or scenario; involves creative leaps.

PREDICTING EMERGENCIES

Hazards are conditions that with exposure present a risk for negative outcome or loss. A simple example is a sharp edge that, when touched in the wrong manner or without protection, can result in a laceration or penetration of the skin. Vulnerabilities are conditions that allow a person to exploit or breach security or gain access to an asset. Threats are different and involve human intent to cause harm. Emergencies are the resulting consequences from hazards, threats, and vulnerabilities and also include natural disasters that result in potential loss. These emergencies are a type of incident that can be unintended, intentional, or naturally occurring that require a timely and proportional response of human and material assets to limit loss. The only proactive preparation is to plan for response and to implement controls in attempt to prevent loss. This description renders definitions of the word "emergency," that include the element of being unexpected, obsolete.

Emergencies are not unexpected. The time and place may come at a surprising time, but our job in planning response to emergencies and preventing them requires us to anticipate them as much as possible. The effort begins with an inventory for hazards, aspects, security inventory, and vulnerability assessments. It continues as an ongoing effort when organizations examine near-miss situations, including examining for reasonable consequences from the results of an incident. Organizations must also stay up to date with trends that surround threats because many criminal acts spawn future copying. In this sense, we are looking to make predictions based on organizational experience and societal experience.

In emergency planning we have categories of emergencies that must be addressed. Natural disasters are the first category. In conducting our initial assessment, we utilized resources to research past occurrences to predict possible natural disasters that could occur on our grounds. In evaluating these natural disasters it also becomes apparent that some of them may not impact our facility from direct exposure, but will impact business operations due to the impact the disaster will have on the associates and their ability to work. Typically, this is handled in business continuation planning, but many times aspects of this type of planning impact are included in the emergency response planning as well. A good example of this is flooding. Large-scale flooding may not harm our facility due to it being located outside of a flood plain, but it could become isolated from the community and the workforce. Issues such as utility services and workforce availability may require special arrangements such as altering shift times, housing personnel in close-by hotels, or performing special shutdown procedures for critical processes. Another good example of an emergency that has similar impact on the facility and business is a breakout of illness or a pandemic. Widespread and serious health impacts of employees' family members can also impact availability and business as much the impact of the illness on employees directly.

By researching area history for natural disasters with the National Weather Service, National Oceanic and Atmospheric Administration, and local or regional dispatch centers, emergency management offices at the local or state level, and with other local industries, can develop a tailored list of natural disasters to plan for. Typical natural disasters can include floods, tornadoes, hurricanes, earthquakes, wild fires, or pandemics. Table 5.1 gives an example of a template for considerations when planning for these disasters.

TABLE 5.1   *Natural Disaster Considerations Template*

NATURAL DISASTER	BASIC STRATEGY	ALTERNATE POLICY	MUSTER POINT
	Shelter in place: Y □ N □	Yes	Indoor or outdoor (circle)
	Evacuation: Y □ N □	No	Away from utilities: Y □ N □
	Active response: Y □ N □	Notes:	Grid coordinates:

Table 5.1 is a good starting point for beginning the planning for a natural disaster. In other words, it serves as an inventory for natural disasters and gives overall guidance when planning, whether by cross-functional committee or the safety department alone. It begins with a basic preference for overall strategy. Will we need to evacuate

or seek shelter indoors? Will we have a team that must respond or perform duties? In some of the natural disasters we may need to consider establishing an alternate policy to normal human resource policy in non-emergency situations. A good example is an alternate attendance policy during outbreaks of illness. The muster point or point where employees will congregate in an emergency is very important and is a critical area of planning.

The muster area or rally point is a place of safety. In an outdoor area, planning considerations must include being away from utilities such as gas lines, being out of the predominant direction of the wind in case of fire or airborne contaminants, having adequate access routes, having access routes that interfere as little as possible with responding agencies, and pre-planning an area for landing airborne medical helicopters. Indoor areas for shelter also have critical planning considerations. They include being away from flammable and combustible materials, having access to water, exit access (to not become entrapped), if possible, being as free as possible from flying debris, the ability to communicate, and structural stability.

Much of this initial preparation facilitates response to threats. Threats originate with human intent. They are largely predicted by past experience and are difficult to pinpoint when, how, or whom will next enact a threat. At the time of writing this chapter, Stephen Paddock had just committed the most heinous mass shooting in the history of the United States, killing fifty-eight people and injuring over 500 more at an outdoor concert in Las Vegas. It is not unpredictable from the standpoint of possibility. It was unpredictable from the standpoint of plausibility. In other words, we balance or manage risk to an acceptable extent based on experience. Up to this point, it was unheard of to fire down on people from a relatively far distance, from a high vantage point, at a large-scale outdoor event. We, as a society, accept inconvenience only so far as our near vision allows. There are valid reasons why this is human behavior. It unduly taxes our resources. Therefore, we plan for threat response on what is plausible from the review of past experiences, which also become timely in nature. For example, bombings became common in the late 60s as some anti-war groups adopted more militant tactics for awareness. In the 90s, suspicious substances in packages, such as an anthrax scare, became a common concern.

Threat response begins with establishing a threat-level communication method tied to increased security measures. Mirroring our national threat-level system, there are times when the potential for a threat incident increases. Examples include times of personnel termination or labor disputes and domestic violence situations that spill over from home and at times of civil unrest. When we perceive an increased risk, we can communicate this by color level and assign additional duties to key associates to increase our ability to detect the beginning of a threat as early as possible. If we receive a confirmed verbal or physical threat, we have a credible threat or one that is imminent and we can increase security measures even further than a lower level or a perceived increased risk. In other words, we can tighten security the more credible the indicator of imminent threat exists. Table 5.2 reflects such a communication plan.

TABLE 5.2 *Communication Plan in Case of Threat*

SECURITY CONDITION	THREAT LEVEL	MEASURES
No perceived increased risk	Green	Normal security practices in place
Probable increased risk	Yellow	*Security stopping all incoming vehicles for ID *Bay doors no higher than five feet for ventilation *Mesh gates behind bay doors locked *Communication sent to all managers and supervisors as to facts behind increased threat level
Imminent threat likely	Orange	All yellow condition measures plus the following: Physical checks of entry points every thirty minutes All cameras working and recording
Threat received or imminent	Red	All previous measures plus the following: *All cameras monitored for detection purposes *All vehicles stopped and checked for validity; only associates and planned deliveries accepted. *Details of persons of interest or possible suspects and descriptions communicated to supervisory and management-level personnel.

Table 5.2 is a minimal example for communicating the point of such a communication system. It may not be perceived to be effective by noninvolved personnel, but it is effective for tightening security to increase the ability for early detection. The exact measures would depend on the resources and security plan of the organization. But it is an example of one situation in an ideal manner. Let's say that a verbal threat of a bomb is taken over the phone by a receptionist. We would treat the threat as credible until a preponderance of the evidence indicated that it was not credible. We may have to evacuate, wait until a specific time passes, possibly search for suspicious items before or after a timed threat or before reentry is made. But then we would tighten security to the point that we can reasonably assure that entry has not been made without our knowledge and that we are free from suspicious items. A subsequent phone threat may not warrant an evacuation if we can guarantee clear conditions.

Common threats for planning include bomb threat, suspicious packages by mail or by recognition, active shooter, crimes against property, actions against persons, or workplace assaults. These threats can originate from internal associates or persons external to the company, either known by an associate or a stranger, and all are violent. The following sections give some general advice about planning for these threats.

THREATS

Bomb Threats

A very common practice is to place a threat recognition sheet at every phone station that has access to an outside line. In the event a person answers a call where a verbal threat is made, the sheet acts as a guide to handling the call and as a record of critical information. Many times these cards or sheets are lost or removed from the phone station. Therefore, make it an attached and standard item that is inspected for routinely.

Standard response is to evacuate to a safe distance, establish search teams that include someone familiar with the area assigned, search for anything out of the ordinary, and stay back until after the designated time (if the threat was a timed threat), and when reentry is made, security is tightened to the extent that a clear area is maintained and guaranteed.

There have been all kinds of debate on possibilities, such as not evacuating the first time because it could be a ploy to move a target into an open and accessible area. The first threat could be a ruse meant to move the targets back inside if the item is missed to maximize casualties. Again, these are possible scenarios and not plausible scenarios based on past experiences. There is a real dilemma in that someone makes the call to expose personnel or take the most conservative route. The conservative route exposes the company to the motive of work interruption. So, the situation depends on the facts surrounding it.

Suspicious Packages

This part of the emergency response plan is akin to the bomb-threat plan. We want to control the places and types of personal items allowed in specific areas, control deliveries, and limit exposure to mailed packages that are receive into the facility.

Personal items such as purses, bags, and lunch boxes should not typically be allowed in production areas. However, in smaller manufacturing operations, warehouses, and such facilities, lockers are not always available. We must identify the areas where these personal items are to be stored. This is a critical piece of planning when searching an area for suspicious packages or items that do not belong. It is also important in limiting exposure to workplace substances.

Deliveries must also be managed in that they are expected and inventoried as they are received. Any item not on a shipping manifest should be quarantined and checked against the manifest and then confirmed with the shipper. Many times materials are packaged together on pallets; the person receiving the material must at least inspect the packaged materials for any of the signs of a suspicious package that are listed in Table 5.3.

Mail is also a concern; even though the post office may scan and screen mailed items, many times a suspicious package is placed in with mail at the box or receiving point. Some companies will utilize a post office box; others will not. We should limit the duty to one person and have that person sort and inspect the mail in an area where indoor ventilation is not present and exposure to other persons is eliminated. The person receiving the mail will look for signs of a suspicious package, as listed in Table 5.3, and, if encountered, should immediately notify the safety department or company contact via non-face-to-face communication and subsequently local law enforcement if the signs are overwhelming or if opened and a substance is encountered.

Table 5.3 contains the signs associated with a suspicious package.

TABLE 5.3 *Suspicious Package Signs*

SYMPTOM	FINDINGS
No mailing address	Y☐ N☐
No return address	Y☐ N☐
No postage/postage markings	Y☐ N☐
Too much postage	Y☐ N☐
Excessive taping	Y☐ N☐
Leaking materials	Y☐ N☐
Odors	Y☐ N☐
Poorly spelled words/addresses	Y☐ N☐
Unusual shapes/bulges	Y☐ N☐
Handwritten addresses combined with previous condition(s)	Y☐ N☐

Source: US Department of Homeland Security 2012.

Active Aggressions

Workplace assaults happen because of human relationships. We begin with policies and training about horseplay and personal conduct in regard to public shows of affection and forms of harassment. The management of the organization must make a commitment to response in case of assault or an ongoing physical altercation. Will the organization commit to getting personnel to stop the altercation and if so, how? Is relying on local law enforcement to stop the altercation enough? Several minutes' response time is a long period of escalation. Serious injuries can occur, actions can get more violent, and typically bystanders will attempt to break up the fight. If personnel are used to respond they will have to be trained and equipped to do so. Regardless of the use of personnel to respond, nonresponders, or all associates, if personnel were not used to respond, would have to be trained on their duty of summoning more advanced aid and on the possible results of trying to stop the aggression. This precedent for training is established in 29 CFR 1903.14(f).

Personnel responding to restore order should be trained in the force continuum acceptable in state laws, which protect citizens in such situations, or Good Samaritan laws; in personal legal liability, techniques for separating altercations, and training on any equipment issued such as pepper spray; and on giving statements to law enforcement once they have responded and restored order. Is this a level of commitment that the organization is willing to allocate resources toward?

In some industries where there is an increased exposure or contact time with the public as a customer, the threat of assault is a direct and foreseeable job hazard. In these situations, separation from the public is the key. The National Institute for Occupational Safety and Health has prepared various publications addressing special industries where contact with the public seems to be a common. In 1996 NIOSH published a set of risk factors that were derived from several studies. These risk factors were as follows:

- Contact with the public
- Exchange of money
- Delivery of passengers, goods, or services
- Mobile workplace, such as a taxi police car
- Working with unstable or violent persons
- Working alone or in small numbers
- Working late night or early morning hours
- Working in high crime areas
- Guarding valuable property or possessions
- Working in community-based settings

Source: Center for Disease Control and Prevention 1996, 14.

NIOSH recommended an abatement strategy that combined environmental designs, administrative controls, and behavior strategies. Environmental design strategies included cash handling strategies, physical separation by bullet-resistant barrier, high amounts of lighting, access and egress points, and personal protective gear. Administrative controls consisted of staffing plans and scheduling, work practices and procedures, and reporting methods for threats. Behavioral strategies made up most of the training around workplaces, such as how to seatbelt a violent person without exposing yourself to a biting threat, verbal de-escalating techniques, and use of personal protective equipment (Centers for Disease Control and Prevention 1996).

Arson

The threat of fire by arson is a reasonable threat, but it may not be discovered immediately in the response. Normal fire-prevention techniques are applicable here because these techniques make them more recognizable. Walk-through inspections of critical areas, such as flammable store rooms and near piles of combustibles, should include inspection for ignition devices and subversive behavior observations. Inspection for lighters, for example, and observation for smoking or for unauthorized cell phone usage not only may prevent an unintended result, but may also point out possible sabotage efforts.

Active Killers

The topic of active shooters or persons actively killing with a weapon other than a firearm, such as a knife or sword, have become popular incidents for training at schools and at manufacturing plants. Training concentrates on recognition of suicidal behavior, reporting importance, sounding alarms, barricading for protection, cover and concealment, and lastly, taking a stand.

Limiting harm depends on the type of barriers and level of penetrability of those barriers in keeping a person separated from associates. But in many manufacturing settings, it is an employee or former employee who either gains entry in a normal manner or utilizes a vulnerability to gain access. Developing entry procedures and training personnel on strict adherence is critical but not that reliable. Let's look at an entry door protected by code or utilizing associate ID cards for activation and unlocking. Even then, an employee may recognize another as being a part of the organization and hold the door open for him or her as a courtesy. Being able to communicate an increased threat level and increase security measures in correlation to the threat level is a possible counter that can help.

A proactive strategy is to perform rigorous background checks on prospective hires. Controversial in nature, but applicable, is the requiring of criminal background checks on current employees prior to promotion, bidding on new positions, and self-reporting of occurrences or reporting on a routine schedule. Commercial drivers, for example, must disclose violations and criminal acts for their driving records. A policy such as this could be used to eliminate persons of a violent nature.

Companies must also stop negligent retention of employees who have committed assaults or crossed the line into a serious crime or potentially injured a fellow worker with horseplay. Another issue of interest is double victimizing a worker who defends his or her self. Zero-tolerance policies can create backlash. If someone defends him- or herself, then disciplining him or her makes that person a victim of the bully and of the company. Victimization is a mental trauma and when a person is doubly wronged, then he or she will feel that he or she has little voice and little understanding and will likely, after experiencing shock, experience anger toward all.

The life-changing units theory of accident causation theorizes that stress increases the potential for a worker to be injured or involved in a workplace incident. The more stress that a person accrues, the more likely they will be involved in a workplace incident. The theory evolved from a medical study of stress and its correlation to life-threatening illnesses. From the study and applied to a workplace setting, the following examples are some of the stressors that increase the potential for mishap:

- Death of spouse
- Divorce
- Personal injury
- Loss of job
- Pregnancy
- Change in financial status
- Change in work duties
- Change in a family member's health
- Outstanding personal achievement
- Change in sleeping habits
- Change of residence
- Vacation (Alkov 1972)

We can use the complete list to educate our supervisors for a recognition program that could then refer or suggest counseling to at-risk workers. At a minimum we could include them in conducting safety reviews or assisting in other safety-related activities to help change their morale and increase their attention toward safety in attempt to prevent involvement in an incident or further degradation of their personal morale or depression.

Supervisors are in the perfect position to observe for these stress events. Supervisors have a closer relationship to employees due to frequent contact and a duty to observe how problems in and out of the workplace affect work performance. Let's begin to look for early signs of employees who are in trouble and attempt to bring them back up to healthy performance levels.

Risk factors that are known to be associated with violence toward others or to one's self are categorized as school-related factors, personal risk factors, communal risk factors, and factors associated with family. It is very important to recognize that that these factors are not direct enough to be considered direct causes but are underlying predictors that must be considered as a whole and as a pattern for risk. Having a mental health partner that can help provide counseling services is an important component to prevention of workplace violence. Table 5.4 reveals the categorized risk factors according to the New York Office of Mental Health (2018).

TABLE 5.4 *Risk Factors for Violence*

INDIVIDUAL RISK FACTORS	History of violent victimization
	Attention deficits, hyperactivity, or learning disorders
	History of early aggressive behavior
	Involvement with drugs, alcohol, or tobacco
	Low IQ
	Poor behavioral control
	Deficits in social cognitive or information-processing abilities
	High emotional distress
	History of treatment for emotional problems
	Antisocial beliefs and attitudes
	Exposure to violence and conflict in the family
FAMILY RISK FACTORS	Authoritarian childrearing attitudes
	Harsh, lax, or inconsistent disciplinary practices
	Low parental involvement
	Low emotional attachment to parents or caregivers
	Low parental education and income
	Parental substance abuse or criminality
	Poor family functioning
	Poor monitoring and supervision of children
PEER AND SOCIAL RISK FACTORS	Association with delinquent peers
	Involvement in gangs
	Social rejection by peers
	Lack of involvement in conventional activities
	Poor academic performance
	Low commitment to school and school failure
COMMUNITY RISK FACTORS	Diminished economic opportunities
	High concentrations of poor residents
	High level of transiency
	High level of family disruption
	Low levels of community participation
	Socially disorganized neighborhoods

Source: Centers for Disease Control and Prevention 2018.

Preventing threats of violence by employees is critical. Early recognition of an employee who is in trouble and referring to counseling is good practice. But there are preventive measures we can borrow and adapt from our studies involving educational settings. Adapted from a study of school practices conducted by Sprague and Walker in 2005, we can do the following:

- Provide adequate job mentoring and professional development
- Implement consistent and just disciplinary practices
- Increase employee participation opportunities and self-management practices
- Set clear expectations and rules of conduct
- Recognize individual differences and practice inclusion
- Offer mental health counseling
- Practice frequent management contact

The New York Office of Mental Health (2018) recommends practicing "protective factors" as a counter to many of the previously listed risk factors. Although written toward dealing with youths, comparable management practices in industry can be interpreted as countering individual characteristics with matching cognitive ability with job requirements, looking for a positive temperament in interviews, and recruiting and looking for social skills and social intelligence in pre-employment processes. Furthermore, the company can formally set and recognize positive behaviors formally and publicize positive actions. Along this line, the company can utilize committee work and problem-solving groups.

UNINTENDED CONSEQUENCES FROM ACTIVITIES

After performing the hazard inventory and mapping for our safety planning assessment, the activities of the organization become clear. Hazards typically have sources, and sources have results or harms that are physical or health related and either acute or chronic in nature. In many cases there are dual threats. Even a laceration carries the possibility of not just physical harm, but of infection, for example. We can use the results of the hazard inventory to predict and prepare for possible incidents that arise from our activities.

Machinery has typical concerns or incident and injury types associated with it. For example, power presses are known to cause amputation and crushing incidents from maintenance activity. The regulations center on the controls for these incidents. Of course, our safety policy and procedure are preventative in nature. However, preparation for such an event can save lives. It also provides an excellent opportunity for bringing in cross-functional groups to perform a table-top exercise and prepare specific response protocols. These protocols can be added to an appendix of the emergency response plan and be disseminated to appropriate personnel in a guidebook. These rescue procedures are expanded on later in this chapter. This is one reason that emergency response planning is an ongoing or living preparation plan that continually grows and improves. It is also a classic example of how the goal of leadership instilling safety as a virtue is facilitated through policy.

Fire is one of the standard unintended consequences that must be planned for. Fire can also be presented as a threat by arson. Preparation begins by identifying possible fuel sources and ignition sources by area and by activity. Dividing the area of the facility by fire-risk rating provides an opportunity for increased awareness. Personnel training and awareness can increase employees' ability to recognize basic hazards. A good example is dust. Even common dust can ignite. For example, overhead mounted fans for circulating air in hot environments can add to the risk presented by fire. Fans will build up dust and when bearings, engines, or other ignition sources such as sparks or slag from welding come into contact with the dust, the fan will erupt in fire and possibly spread. Do workforce-level associates find buildup of combustibles and especially dust as a basic hazard? Do daily workstation clean-up procedures involve cleaning overhead mounted fans? Does the facility have a dust-control plan?

As the hazard inventory is completed and specific sources of hazards are identified, we can, as a piece of our fire-prevention plan, identify specific sources of the chemical category (fire is a chemical reaction) that increase the risk of fire as a type of hazard, rate the risk for fire by location, and suggest management practices at an appropriate frequency. This will also help us plan for portable fire extinguisher placement and use, covered later in the chapter. Table 5.5 suggests a format for such a planning method.

TABLE 5.5 *Fire Extinguisher Placement Plan*

CATEGORY	TYPE/ CLASS	SOURCE(S)	PLANT LOCATION	LOCATION RATING
Chemical reaction	Fire			H☐ M☐ L☐

Outdoor areas cannot be overlooked for fire risk either. In the western regions, planning is standard. But in all areas, landscaping and build-up of vegetation can contribute. For example, mulch around flower or shrubs can ignite, especially when it is dry and an employee or passing motorist discards a cigarette. Control of vegetation and decorative vegetation and materials is a concern at any location.

REFLECTION 5.1

1. What is the key difference between threats and natural disasters and unintended consequences?
2. Besides examining equipment and activity, what other sources may provide insight into unintended consequences that need preparation?
3. Do you think that awareness of possible unintended consequences is increased as procedures are developed and rehearsed? Do you think that as awareness increases, potential occurrence decreases?
4. How will you decide what information in the plan will be accessible to all personnel?

GENERAL PLANNING

Planning begins with identifying the buildings and ground entry points to ease communication between company personnel and between company personnel and visitors, vendors, customers, and even first responders in an emergency situation. Clear signage must accompany entry points from the first access at the property's edge, and signs must be located at specific points to direct incoming persons to the correct destination point. For example, a master sign for multiple entry gates may direct deliveries to one gate and pick-ups from shipping to another while directing visitors and guests to the front gate. Subsequently, at each gate, the gate must be labeled, its use disclosed, and, if not readily apparent, the route must have additional directional signage.

All possible access points to any building must be identified. Tactical first responders in law enforcement use a variance of a specific tactic. It would be good practice to utilize this technique and to visibly identify doors in this manner so that interagency communication is easy. The technique begins by numbering the sides of any building. The front, or typical main entrance, is side one. The sides will be numbered in succession from left to right. Figure 5.1 indicates this technique.

From the left we will number the doors by type and in succession for each side. Door types are typically categorized as an entry door or a bay door. So, while looking directly at a side from the left we can identify each entry-type door as E for entry and label them as 1, 2, 3, etc. for as many entry doors as the building possesses. Bay or roll-up style doors can be identified as B for bay door and further identified by successive number. B-3, for example, would be for bay door 3 on any side. The first entry-type door on side I would be identified as 1, E-1. The first entry-type door on side III would be identified as 3, E-1.

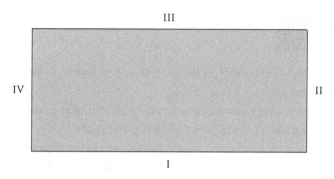

FIGURE 5.1 Building Side Numbering

Windows follow the same identification steps with the difference being that windows are identified with a "W" for window. We can get more complex with varying levels or floors. For example, a building that is two stories would be handled by identifying the floor along with the side. The second story of side I would be communicated as side I, 2. The second story on side II of the building would be side II, 2.

Labeling access points to your facility in this manner allows everyone to focus on a common point with less description being communicated. Different agencies may use different methods, and it may be a good idea to coordinate this effort with them. However, tactical personnel practice and perform their techniques to the point that different identification schemes should not interfere with their response.

POSITIONAL DUTIES

All programs have three levels of responsibility and the emergency response plan is not an exception for duties. All non-rescuers have the duty to recognize emergencies, report emergencies or sound alarms, and follow established procedures or perform as directed by a competent-level rescuer. Rescuers are divided into the authorized level, competent level, and administrator level.

Employers have the duty to ensure a first aid–level response within a reasonable timeframe, which, according to the Brogan letter, is interpreted to mean four minutes (OSHA, Letter of Interpretation 2007). If a medical facility or first responders are not within or cannot meet this timeframe, the employer must meet its burden. Therefore, an authorized rescuer must at least be trained in first aid–level response at minimum and should be certified or have evidenced equivalent training. Rescuer duties are organized with corresponding competency levels in Table 5.6.

In addition, there may be specific positions required by operational activities or by OSHA requirement that must be fulfilled. For example, we will need someone to be a driveway tender, or person who directs responders physically toward the direct route of entry. We will have some designated as evacuators, who assist in and along the evacuation route for employees. Evacuators may help with an injured person but may also stop traffic or guide confused persons. Some of our maintenance or engineering personnel will be utilized to shut down special processes and utilities such as natural gas, or to respond to the alarm panel to confirm the emergency if it is not readily apparent. All these personnel must have radios to communicate with each other. Table 5.6 lists suggested personnel positions.

TABLE 5.6 *Personnel Emergency Positions*

LEVEL	SPECIFIC POSITION	COMPETENCY	DUTIES
Authorized	Driveway tender	First aid training Emergency action training Basic traffic directing safety	• Guide responding agencies to correct entry points • Identify locations of hydrants/ resources
Authorized	Evacuator	First aid training Emergency action training Basic traffic directing safety	• Guide associates to muster point • Assist those injured/confused • Answer/assist employees with questions on the plan
Authorized	Alarm panel monitor	Engineering/maintenance personnel Electrical authorized Authorized LOTO person First aid training Emergency action training	• Respond to alarm panel to confirm alarm and emergency • Coordinate emergency response equipment/alarm inspections
Authorized	Gas/utility shut off (times 2)	Engineering/maintenance personnel Electrical authorized Authorized LOTO person First aid training Emergency action training	• Respond to utility control point to shut down utility.

Continued

Authorized	Critical process shutdown	Certified operator personnel Electrical authorized Authorized LOTO person First aid training Emergency action training	• Remain behind if possible to shutdown critical process
Competent	Public relations	Emergency action plan training Experience/knowledge of operations Knowledge of area media outlets	• Assist in writing public information announcements with confirmation with top executive and/or general counsel to disseminate public information
Competent	Business continuation liaison	Emergency action plan training Business continuation plan training Experience in operations	• Begins putting the plan into action for continuing business and minimalizing damage from disruption
Competent	Operations director	Emergency action plan training Experience in operations Familiar with most personnel	• Coordinate roll call • Report to senior executive • Assist in business continuation • Direct personnel actions
Competent	Emergency response director	Emergency action plan experience Familiarity with responding agencies Competent/administrator safety plan	• Act as liaison with incident command • Direct company responders
Administrator	Emergency response director	Experience scene/behavior analyst	• Prepare and lead after-action reviews • Direct continual improvement

In an emergency, access to the emergency response plan and all appendices such as maps and company documents is a concern. Today, electronic versions or access to a digital database that is password protected can be useful. But emergency managers and responders always have redundant backup plans. Therefore, it is a good practice to place or require certain upper-level managers to keep hard copies secured in their vehicles to have copies if other means of access are not available. Placing a weather-tight lockable box with all plans and necessary items for supply in a safe place to access outside of the building or immediate site area can also take the place of individuals being responsible for carrying a secured copy in the vehicles.

Not all the emergency response plan data should be disseminated to all employees or made public. Information that is of a sensitive nature, such as business continuation data, hazardous material location areas, alarms and detection capabilities, or asset location that may be in demand or be an antecedent to a theft should not be made available to all personnel or the public. It is common to have public plans that cover OSHA-required minimums and then private, more detailed plans.

REFLECTION 5.2

1. How does the identification of possible entry points on a building facilitate emergency response?
2. What other duties would you assign and to what level of competency?

GENERAL REGULATORY REQUIREMENTS OF EMERGENCY ACTION PLANNING

In general, 1910.38 of 29 CFR cover the minimum elements to be included in an emergency action plan. These plans typically run into or merge with other plans and programs. I prefer to include these other plans or portions thereof directly into the emergency action plan as well. For instance, hazardous material storage, handling, and disposal plans are also directly relevant to the emergency response plan. Another good example is the investigations management plan. Portions of the investigations plan, such as handling a scene, are also directly applicable in emergency response. Placing redundant plans in the emergency response plan facilitates the exchange of information rather than referring the reader to another document.

The minimum elements for an emergency action plan are as follows:

- 1910.38(C)(1): Procedures for reporting a fire or other emergency
- 1910.38(C)(2): Procedures for emergency evacuation
  - o Type of evacuation
  - o Evacuation route assignments
- 1910.38(c)(3): Procedures to be followed by employees remaining behind to shut down or operate critical plant operations prior to evacuation
- 1910.38(c)(4): Procedures to account for all employees after evacuation
- 1910.38(c)(5): Procedures for employees performing rescue or medical duties

- 1910.38(c)(6): A list of the names or job titles and contact information for employees that may be contacted by other employees with questions about the emergency action plan and their duties
- 1910.38(d): A description of the employee alarm system ensuring a distinctly different sound for different emergencies
- 1910.38(e): Employees must be trained in orderly evacuation
- 1910.38(f): The employer must review the action plan with each employee covered by the plan when the plan is developed and the employee assigned and when the employee's responsibility changes or the plan is changed (29 CFR 1910.38).

Emergency action plans should also have elements required by safety management system standards. Beyond the legal requirements of 1910.38, safety management systems will require periodic tests, drills, and exercises to be conducted, where practical, after-action reviews will be needed for continual improvement and communication of basic plan elements, as appropriate with peripheral customer of safety, such as contractors, vendors, ground guests, local community entities, and government authorities, will also be needed.

After-action reviews are important for continual improvement efforts in emergency action planning. After any drill, exercise, or emergency, the administrator of the program, typically the safety manager, should prepare an after-action report and lead a cross-functional committee with employee representation to review strengths and weaknesses and produce countermeasures for improvements.

Drills for new programs are progressive and should be graded or assessed. A level 1 drill is an announced and planned drill with the purpose of demonstrating that all associates can demonstrate their duty under the plan. It is not a surprise and reinforces initial training. Competent-level positions, plus other select associates, can be used to assess how well the plan is carried out. For example, department managers can time and assess the evacuation of their personnel, and engineering personnel can assess critical operations shutdown. Afterward, the administrator can cumulate the results and prepare an after-action report and present the findings to the cross-functional committee, and the committee can produce documented improvements.

A level 2 drill is partially announced. This time, only those competent-level personnel who will assess performance know ahead of time about the drill. This assessment may uncover additional problems when a startle response is present. A level 3 drill is only known to critical personnel, and the knowledge of the drill is not spread beyond minimum critical personnel. It can still be graded if outside personnel are utilized to grade the drill.

Exercises can also have varying levels of complexity. Exercises typically involve outside agencies or personnel. A level 1 exercise would be an in-house table-top consideration of scenarios but will still involve outside personnel to advise their agency's role and probable response data and to also assess for weaknesses in the plan and response. Level 2 exercises may involve practical scenario role playing. But exercises are again reviewed for improvement.

Impairment procedures are vital. During drills and exercises, proper impairment of detection, alarms, and equipment must be practiced. This equipment will be placed in a functional but deactivated mode. Alarm-monitoring services will be notified prior to any drill or exercise and be reactivated once the drill or exercise has concluded. Regardless of the monitoring service being notified, local emergency first-responding agencies should be notified centrally of the occurrence so that they do not respond erroneously and waste time or resources that could be needed in actual emergencies. Dispatch centers will call and confirm, with the administrator level of the emergency action plan, any call for service during the times of the drill or exercise.

Impairment is not just for drills and exercises. Impairment procedures are to be written and included in any emergency action plan. If any part or section of detection, alarms, or equipment must be placed out of service for repair, upgrade, or testing, not only is notification to the monitoring service and local agencies practiced, but all cross-functional departments should be notified of the occurrence and the location should be labeled as such to prevent hot work or any activity that increases the chance of fire or other emergency from occurring during impairment.

GENERAL REQUIREMENTS FOR FIRE PREVENTION PLANS

The minimum requirements of a fire prevention plan are contained in 29 CFR 1910.39. Per OSHA, fire-prevention plans are only required when another standard applies to the activity of the organization. These applications are within 1910.157, which addresses portable fire extinguishers; 1910.1047, which addresses exposure to ethylene oxide; 1910.1050, which addresses exposures to methylenedianiline; and 1910.1051, which addresses exposure to butadiene. 1910.157 serves as a catch-all application for the requirement of fire-prevention plans. Fire-prevention plans are a necessary component of emergency response and prevention plans.

Regardless of an exemption for a written plan for employers with fewer than ten employees, fire-prevention plans should be written, even if merely used to document and prove that covering the plan with employees has been completed.

Paragraph c of 1910.39 covers the requirements as listed:

- 1910.39(c)(1): List of all fire hazards, proper handling and storage procedures for all hazardous materials, potential ignition sources and their control, and the type of fire protection equipment necessary to control each major hazard
- 1910.39(c)(2): Procedures to control accumulations of flammable and combustible waste materials

- 1910.39(c)(3): Procedures for the regular maintenance or safeguards installed on heat-producing equipment to prevent accidental ignition of combustible materials
- 1910.39(c)(4): The name or job title of personnel responsible for maintaining equipment to prevent or control sources of ignition of fires
- 1910.39(c)(5): The name or job title of employees responsible for the control of fuel sources hazards

From the requirements contained in paragraph c, a combined template for this plan is suggested in the box below:

Area of plant/site:

Date of review: Next review:

Fire hazard type   Hazardous material name:

Hazardous material ☐ Heat-producing Equipment ☐ Combustible ☐ Flammable ☐	Storage/handling:

Ignition sources: Y ☐ N ☐

Source name/location:	Controls:	Inspection checks:	Responsible person:

Heat-producing equipment: Y ☐ N ☐

Name: Fuel:	Location:

Control:	Inspection/maintenance checks	Responsible person

Housekeeping procedures:

Dust control: Y ☐ N ☐	Schedule/rate	Responsible person

Waste removal: Y ☐ N ☐	Schedule/rate	Responsible person

Fire-protection equipment:

Hazard characteristics:	Fire-protection equipment:

Fire-protection equipment inspection schedule:	Responsible person:

This plan can be organized by plant or worksite area since there may be multiple entries per location. It helps in organizing the data per area of the plant rather than maintaining a master list of multiple items located in various places. It also may help in organizing the responsible person per plant area if maintenance personnel specialize in certain machinery or plant areas. Application of listing hazardous materials by plant area is why hazardous materials are inventoried with plant locations during the hazard inventory process.

Most areas will be considered in the dust-control plan, and all locations have areas for accumulating and discarding waste. These are all part of fire prevention. Controls on heat-producing equipment and other ignition sources can be shown by temperature controls and fuses or alarms tied to temperature levels. Controls will always have an inspection rate and need general checks completed by a responsible person. Many times this is handled and scheduled through the maintenance department's method for issuing work orders. Keeping work-order copies for items tracked in the fire-prevention plan becomes evidence of maintaining controls for heat-producing equipment and ignition sources. Equally important but not required of 1910.39 is the inspection and responsible party for checking fire-protection equipment.

PLANNING FOR EVACUATION

There are several other legal requirements for emergency response planning that originate from the general industry regulations contained in Part 1910. Exit routes are covered in 1910.36, for example. Some basic terms must be understood prior to exit-route planning. An exit route is a continuous and unobstructed path of travel from any point in the workplace to a place of safety. It has three parts: the exit access, the exit, and the exit discharge. Exit access means the portion of an exit route that leads to a properly separated and fire-resistant, rated, and protected area or exit. The exit then is an area that is separate and constructed of proper fire-resistant materials that leads to an exit discharge. An exit discharge means the part of the exit route that leads directly outside or to a street, walkway, refuge area, public way, or other open space with access to the outside (29 CFR 1910.34).

The list that follows reveals requirements associated with design and construction for exit routes contained in 1910.36:

- Exit route design is permanent.
- Number of exit routes is adequate considering the number of employees, size of building, occupancy, and workplace arrangement, with a minimum of two exit routes separated so that if one is blocked the other is open. Single routes are allowed when all employees could evacuate safely in an emergency.

- Separate exits are needed from workplace areas with materials of proper fire-resistance ratings per the number of building stories the route connects. Materials must be a one-hour fire rating when connecting three or fewer stories, and a two- our rating when connecting four or more stories.
- Fire doors and their hardware must be approved by a nationally recognized testing laboratory.
- The outdoor area accessed by exit discharge must be large enough to accommodate all personnel assigned or likely to use that exit route.
- Direction of travel must be clear.
- Exit doors must be unlocked and capable of being opened from the inside without any special tool, key, or knowledge. There is an exception for correctional facilities and facilities of mental and penal activities, when supervisory personnel are on continuous duty.
- A side-hinged door must be used that swings in the direction of exit travel.
- Exit-route capacity must support the maximum occupant load. Exit-route capacity may not decrease in the direction of exit travel.
- The width of an exit route must be wide enough to accommodate the maximum permitted occupant load.
- Ceiling height must be at least 7 feet, 6 inches tall and any downward projection cannot go below 6 feet, 8 inches from the floor.

Determining the occupant load begins with classifying the occupancy and hazard of contents. NFPA categorizes occupancy as assembly, educational, daycare, health care, detention and correction, residential, lodging, mercantile, business, industrial, and storage. Each can have further classification. Hazard contents are described as low, ordinary, and high hazard (NFPA Life Safety Code). Occupancy and content hazard determine fire ratings of materials to separate different occupancies and protect exit routes. We must also reference occupancy and occupancy load factors to determine occupant load and subsequently plan exit routes to ensure that the route can handle the capacity of persons it serves.

We begin by examining Table 7.3.1.2 of the NFPA Life Safety Code and determining the load factor based on occupancy use. For example, from the table the load factor for industrial use with general and high hazard contents has a load factor of 100. This is the minimum square footage allotted to each person located within this occupancy. For every 10-foot by 10-foot square, one person can be stationed safely in an industrial-use situation with general and high-hazard contents (NFPA 2006). Once we have determined the maximum safe capacity of the work area, we can now look at possible exit routes.

We must ensure that if an exit route has hallways, corridors, or is not open access, the materials that separate the route from a work area meet the fire resistance requirements determined from chapter 6 of the Life Safety Code, specifically Table 6.1.14.4.1 (NFPA 2006).

Travel distance and discharge capacity are the next concerns. We must ensure that the area around an exit be within a specified travel distance and be wide enough to handle the number of persons in the area served.

Distance is to be measured on the floor along the centerline of travel, from the furthest point of the occupancy area to the center of the doorway or exit, going around corners and other obstructions by at least a 12-inch clearance (NFPA 2006, 7.6.1). We must check the appropriate chapter of the Life Safety Code according to occupancy to decide maximum travel distance to an exit. Table 40.2.6 references industrial occupancies, and the maximum feet when the area and route are protected by appropriate sprinklers is 250 for industrial occupancies with other than high-hazard contents (NFPA 2006). In chapter 40 it also specified that industrial equipment access items that are used in a means of egress must be at least 22 inches wide (NFPA 2006 40.2.5.2.1). Therefore, all aisles should adhere to this requirement as a minimum. Doors must at a minimum provide 32 inches of clear space of travel.

When planning or assessing current exit routes, it is clear that utilizing the Life Safety Code is required for basic planning to assess the various possibilities of unique situations.

Further requirements for exit-route maintenance, safeguards, and operational features from 1910.37 are as follows:

- Employees must not travel toward high-hazard areas unless they are shielded; danger must be minimized.
- Exit routes must be clear and unobstructed.
- Safeguards such as sprinklers must be maintained and in working order.
- Exit routes must be adequately marked and lighted so that a person with normal vision can see along the route.
- Each exit must be clearly visible and marked by a sign reading "exit."
- If direction of travel is not readily apparent, signage along the route must be posted.
- The line of sight to an exit sign must be clearly visible at all times or from any point.
- Any doorway along an exit route that is not an exit must be clearly marked "Not an exit."
- Each exit sign must be illuminated to 54 lux by a reliable power source. These are usually battery backup units. Self-luminous signs are permitted if they have a minimum luminance value of .06 foot lamberts.
- Exit signs must be plainly legible with letters at least 6 inches in height and 3/4 inches in width.
- An employee alarm system must be operable.

The Life Safety Code contains some important guidance about the maintenance of means of egress. Of very important notice is that walking surfaces are to be nominally level. Abrupt changes in elevation levels cannot exceed1/4 of an inch. Changes in elevation between 1/4 of an inch and 1/2 of an inch are to be beveled. A change of 1/2 an inch is to be treated like a step and must be marked to be readily apparent or ramped. The slope of a walking surface in the direction of travel should not exceed 1 in 20, unless a ramp is utilized. The perpendicular slope to the direction of travel shall not exceed 1 in 48 (NFPA 2006). These are very important aspects in regard to access of persons with varying movability. Those with vision impairments, walking impairments, and the like need substantially level exit routes. With today's aging and diverse workforce, accessibility issues have to be considered in exit-route planning.

MEDICAL SERVICES AND FIRST AID

The employer is mandated to ensure the ready availability of medical personnel for advice and consultation on matters of plant health by 1910.151(a). This hints at a medical partnership. Medical partnerships are beneficial to the employer in regard to performing medical-related duties that might include, among other things, pre-employment physicals, ergonomic baseline assessments, and follow-up exams, as well as partnering in ergonomic early intervention programs and treatment, hearing conservation programs, conducting drug screenings, and assisting with light duty assignments.

The big issue in emergency response training in regard to 1910.151 is the mandate to provide first aid-level response and supplies. 1910.151(b) states that in the absence of an infirmary, clinic, or hospital in near proximity to the workplace, a person or persons shall be adequately trained to render first aid. Adequate first-aid supplies shall be readily available.

An assessment of the type of injuries, frequency of each type, and severity should be conducted. It begins by reviewing the worker's compensation carrier's report or log of claims. Next, the OSHA 300 log can be reviewed, but realize it will only indicate those injuries or cases that led to medical treatment beyond first aid. Review any injury logs or documents of company history to have an idea of what has occurred in the past. The hazard inventory provides much valuable information as to the view of what possible injuries could occur. By documenting the sources of review, you provide a valid base for establishing what first-aid supplies need to be made available. Table 5.7 may reflect a format for documenting the sources and findings of such a review.

TABLE 5.7   *Workplace Injury Log Example*

INJURY TYPE	FREQUENCY	SEVERITY	SOURCE

The Brogan letter, dated January 16, 2007, established that near proximity or a reasonable response time for first aid–level response should be three to four minutes (OSHA 2007). This confirms that in most situations first aid–level response teams must be established. There must be an adequate number of personnel with adequate first-aid training to provide three- to four-minute response times in emergencies. No medical equipment or supplies can be made available without adequate controls or training to use the equipment or supplies.

Because of the liability created by supplying equipment or supplies without ensuring adequate training and controls, it truly becomes necessary to establish trauma stations around the facility in key locations to provide quick but controlled access. These stations are best located at supervisory stations. The equipment that is included depends on the level of training to response personnel. For example, trained EMTs or paramedics are capable of using much higher level of medical equipment and supplies. But even if first aid-level response is all that is prepared for, not all personnel can or should have access to equipment that may be needed to treat at the first aid level, extricate from the immediate scene, or provide CPR or AED (automated external defibrillator) response. But this type of equipment should not be made available in first-aid boxes that are not controlled.

For these reasons, I recommend placement of many first-aid boxes throughout your facility with basic items such as Band Aids, gauze, Q-tips, ointments, protective gloves, tweezers, and other basic items that your injury review shows a need for. All personnel can access and use these supplies on a personal basis, although only trained first-aid responders should deliver treatment to another person. Trauma kits may contain items such as clotting solutions, trauma bandages, slings, splints, litter cots, CPR masks, AED devices, and any devices or supplies that require special training. These trauma kits allow access to needed equipment by stationing them throughout the facility while not having to expend resources to place them everywhere. But it also controls access to those who are trained.

First-aid responders now have occupational exposure to blood-borne pathogens. Occupational exposure is defined as a reasonably anticipated contact with blood or other potentially infectious materials from the performance of an employee's duties (29 CFR 1910.1030(b)). Because we have trained and assigned first aid-level response to certain associates, they now have an employment duty to render aid. We must develop an exposure control plan meeting 29 CFR 1910.1030. First-aid certification programs cover personal protection and decontamination, but you should also cover disposal of used first-aid supplies and clean-up of any bodily fluids. Furthermore, these personnel must also be offered vaccination against hepatitis B along with any other associates who have occupational exposure.

ESTABLISHING RESCUE DUTIES

Our activities and machinery determine what possible scenarios could occur that necessitate procedures for rescue to provide first aid–level response. Work at heights will always demand preparation for rescuing or retrieving personnel who fall, for example. Special machinery such as presses may require rescue from caught-in scenarios; confined spaces and permit-required confined spaces require preparation for rescue, as well as a host of possible unintended consequences of our activity. The presence of controls does not guarantee the absence of an unintended consequence, unless eliminated a hazard is merely balanced.

First aid-level responders become our rescue team. Like with any managed safety program, we have the authorized level, competent level, and administrator level. Continuing from the breakdown of positional duties from earlier in the general planning section of this chapter, we can establish different levels of rescuers. This is especially helpful when other compliance plans such as fall protection and prevention require planning of rescue and first aid in the event of a fall. So, there are compliance-level programs such as fall protection and prevention that will require these levels to be established and included in the compliance plan. Another critical example is when working with electrical safety compliance and when troubleshooting or live work is conducted and arc flash protection planning is necessary. These levels can be established for each program. However, in general industry, maintenance and engineering positions will crossover multiple program applications. Table 5.8 identifies the basic duties and competency levels for rescuers.

TABLE 5.8 *Duties and Competencies for Rescuers*

LEVEL	DUTIES	COMPETENCY
Authorized rescuer	Recognize the hazard type Report the hazard type Perform basic rescue Maintain/use any equipment Follow procedure as directed	Trained in hazard recognition Trained in basic rescue Trained in equipment inspection
Competent rescuer	Plan basic rescue Inspect equipment Assist investigations Enforce policy/procedure	Trained in rescue plans Trained in equipment inspection Trained in investigations Trained in policy/procedure

Administrator	Overall plan approval	Competent investigator
	Leads investigations	Capable of program tracking
	Overall program performance	

Due to the nature of rescue, depending on the situation and hazards dealt with, in planning for basic rescue per specific compliance program or specialized activity, such as power presses, it is clear that the work load of competent personnel and administrators may have to be shared among other key associates.

To limit liability to citations in the area of emergency response and rescue, we must follow the guidance from 29 CFR 1903.14(f). This paragraph covers the issuing of citations in respect to rescue activities undertaken by an employee in response to an individual in imminent danger. Here, it is clear that the employer has the duty to train and provide for the health and safety of employees who have rescue duties, including providing necessary equipment.

The paragraph goes on to establish that employees not assigned rescue or response duties must be instructed that their duty is not to rescue but to perform other duties such as sounding alarms and summoning more advanced help. Furthermore, they must be instructed in the hazards of performing such a rescue without proper training and equipment. This covers an employer when non-rescuer associates decide to make a decision and attempt rescue, such as jumping into a collapsed trench to aid a fellow worker.

So, the question becomes, "Why plan for rescue and assign personnel?" Of course the four-minute rule established by the Brogan letter begins the answer, but rescue is sometimes and arguably above first aid-level response (OSHA 2007). But 1903.14(3) (i) has a catch-all situation that would apply to many places of employment across many industries. If an employee is in a workplace where it is reasonably foreseeable that a life-threatening accident could occur, the employer has a duty to instruct all non-rescuer personnel in their duties and consequences of attempting rescue without proper training and equipment. The moral implication to not prepare for rescue to the point of being safe to conduct first aid, while educating employees to not render immediate rescue, will have massive negative repercussions on establishing safety as a virtue. It is also arguably in violation of not rendering at least first aid–level response within a reasonable timeframe. For example, if an employee had an arm stuck in an industrial machine, first aid could not be safely rendered until, in many circumstances, the employee was freed from the machine.

PORTABLE FIRE EXTINGUISHERS

Portable fire extinguishers are a critical component of emergency response. They can prevent a minor fire from becoming a large loss and can facilitate evacuation of personnel. 29 CFR 1910.157 establishes legal requirements for the placement, use, maintenance, and testing of portable fire extinguishers.

The first general requirement is to mount, locate, and identify portable fire extinguishers so that they are readily accessible and do not subject employees to possible injury. The National Fire Protection Administration recommends locating extinguishers near paths of travel, close to and along exit routes and at exits, where temporary blockage does not occur. Mounting extinguishers so that they do not fall and create a danger of injury is also important. Mounts must be to walls or columns or in cabinets or recesses and be firmly secured. Operating instructions must face outward toward the user. Mounting height is based on weight, as indicated in Table 5.9:

TABLE 5.9 *Fire Extinguisher Suggested Mounting Height*

EXTINGUISHER WEIGHT	HEIGHT ABOVE FLOOR
Less than or equal to 40 pounds	Top not to exceed 60 inches
Greater than 40 pounds	Top not to exceed 42 inches
No case	Bottom no closer than 4 inches from floor

Source: Conroy, 2008.

Typically fire extinguishers placed for employee use are type ABC, or designed for class A combustible pile fires, type B combustible liquids, and type C electrical fires. This helps alleviate proper selection when stress is high. However, depending on the size of combustible pile or flammable liquids present, the area classified as low-moderate or high-hazard occupancy could be better served by having only type A or type B extinguishers available so that the rating of the extinguisher expressed in pounds or class rating may be adequate to cover the space or square footage. Special applications for hazard such as class D fires (combustible metals) or others are located as to control access and have extinguishers matched specifically to the hazard located in that immediate area only. The National Fire Protection Administration establishes the maximum walking distance per class of extinguisher and the maximum square footage that an extinguisher by class rating can cover. The maximum walking distance for a class A extinguisher is 75 feet, while for a class B extinguisher it is 50 feet. Referring to Table 17.5.3 of NFPA 10 will help you establish the placement of extinguishers. For example, the table reflects that a 10-A-rated extinguisher can cover a maximum of 10,000 square feet of high-hazard occupancy area (Conroy 2008). Class C or ABC class extinguishers are to be deployed on the basis of the existing class A or B hazard (1910.157(d)(5).

Other major requirements for 1910.157 can be listed as the following:

- Fire extinguishers must be approved.
- Fire extinguishers must be maintained with monthly and annual visual inspections, and all documentation must be kept.
- Employees must be trained on extinguishers and other equipment usage at initial appointment and annually.

REFLECTION 5.3

1. If you were re-writing 1910.38, what new requirements would you include?
2. How does development of rescue duties drive safety culture?

OTHER SPECIFIC SITUATIONS FOR EMERGENCY RESPONSE PLANNING

Depending on what the initial hazard inventory revealed, there are situations reflected in Part 1910 that require specific emergency response planning. For example, 1910.119 requires that emergency shutdown actions and operations be incorporated into the written operating procedures for each process. These emergency shutdown procedures would also be reflected in the emergency response plan.

The list that follows reflects the specific situations reflected in Part 1910:

- 1910.66: Powered platforms for building maintenance
- 1910.111: Storage and handling of anhydrous ammonia
- 1910.119: Safety management of highly hazardous chemicals process
- 1910.120: Hazardous waste operations and emergency response
- 1910.124: General requirements for dipping and coating operations
- 1910.146: Permit-required confined spaces
- 1910.156: Fire brigades
- 1910.262: Textiles
- 1910.266: Logging operations
- 1910.268: Telecommunications
- 1910.2699: Electric power generation, transmission, and distribution
- 1910.272: Grain-handling facilities
- Commercial diving operations in Subpart T
- 1910.1003 carcinogens:
  o Alpha-Napthylamine
  o Methyl chloromethyl ether

- o 3,3'-Dichlorobenzidine
- o bis(chloromethyl) ether
- o Beta-Napthylamine
- o Benzidine
- o 4-Aminodiphenol
- o Ethyleneimine
- o Beta-Propiolactone
- o 2-Acetylaminoflourene
- o 4-Dimethylaminoazobenzene
- o N-Nitrosodimethylamine
- 1910.1017: Vinyl chloride
- 1910.1027: Cadmium
- 1910.1028: Benzene
- 1910.1029: Coke oven emissions
- 1910.1044: 1,2-dibromo-3-chloropropane
- 1910.1045: Acrylonitrile
- 1910.1047: Ethylene oxide
- 1910.1048: Formaldehyde
- 1910.1050: Methylenedianiline
- 1910.1051: 1,3-Butadiene
- 1910.1052: Methylene chloride
- 1910.1450: Occupational exposure to hazardous chemicals in laboratories (OSHA 2004).

Activities covered by these regulations may have specific emergency procedures, detection methods, or planning elements that must be accounted for.

EPA EMERGENCY PLANNING REQUIREMENTS

In 1986 the Emergency Planning and Community Right to Know Act was passed and became Chapter 116 of Title 42 of US Code. The legislation aimed at providing public access to information about the hazardous materials present in a community and coordinating this information to formulate local response plans to hazardous materials emergencies such as spills or releases. It accomplished by establishing local and state emergency planning boards and requiring facilities to disclose the presence of hazardous chemicals and the effects of the chemicals and releases (von Oppenfeld 2011).

The structure covered in Subchapter 1 of the EPCRA establishes a state-level emergency response commission (SEPC) that designates districts throughout its state and appoints members of a local emergency planning commission (LEPC) for that district. Members of the LEPC shall include at a minimum the following:

- Elected state and local officials
- Law enforcement
- Civil defense
- Firefighters
- First aid responders
- Health
- Local environmental
- Community groups
- Owner/operators of facilities subject to the legislation (42 USC Section 11001(b))

This means that the facility must have a representative on the LEPC. This will likely be the emergency response plan administrator or safety manager.

A facility is covered under the legislation if it "produces, uses, or stores" a substance in the amount of the threshold planning quantity of a substance listed on the list of "extremely hazardous substances" contained as appendix a to 40 CFR Part 355, or it is covered if it possesses chemicals that require a safety data sheet under 29 CFR 1910.1200. These substances are listed in pounds. Therefore, the safety manager preparing an emergency response plan must have any of these substances cross listed to the unit of measure commonly encountered at the facility to identify whether the substance is present in planning threshold amount, but this will also facilitate reporting requirements and designation of incidental or reportable amounts if a release occurs.

Specific gravity is the ratio of density of a substance to water. Water has a specific gravity of 1 and it weighs 8.34 pounds per gallon. We can then multiply 8.34 times the specific gravity of the substance.

So, let's say that a facility uses formaldehyde in an operation. Formaldehyde is on the extremely hazardous substances list with a planning threshold quantity of five hundred pounds. Therefore, if the facility has five hundred pounds or more of formaldehyde, it falls under the requirements of EPCRA. The facility likely deals with formaldehyde in gallons, so we must know how much a gallon weighs to see if we possess the threshold planning amount. From the safety data sheet we can obtain a specific gravity; for formaldehyde it is 1.1. We can now calculate that it weighs 9.174 pounds per gallon by multiplying 8.34 by 1.1. If we divide 500 by 9.174, we now know how many gallons it takes to equal five hundred pounds. This answer is 54.5 gallons.

In our emergency response plan we could utilize the following form from Table 5.10 to calculate and identify any substance that we produce, use, or store from the extremely hazardous substance list. This information will be used to disclose the location and amount of the substance on a report to the LEPC.

TABLE 5.10   *Hazardous Substance Identification Form*

SUBSTANCE FROM EHS LIST	TOTAL MAX AMOUNT /POUNDS	LOCATIONS/ AMOUNT	THRESHOLDS: POUNDS/ COMMON MEASURE
Name: CAS: Specific gravity:			Planning: Reporting:

REPORTING UNDER THE EPCRA

Any facility that produces, uses, or stores a chemical that falls under the requirements of 29 CFR 1910.1200, the hazard communication standard, must also submit a list of those chemicals for each chemical with the specific information included: chemical name or common name, and any hazardous component of such chemical, an estimate of the maximum amount of hazardous chemicals present at any time, and an estimate of the average daily amount present. This would cover any special mixture. The list must be provided to the LEPC, SEPC, and the fire department with jurisdiction over the facility. This list must be updated whenever a new chemical is brought into the facility or there is a change to a chemical on the list. This list is referred to as a tier 1 inventory and is based on the previous year's report (42 USC 11021). The tier 1 inventory report form queries the chemicals based on a division of hazards categorized as physical or health classes. So, each chemical is not covered separately but grouped per hazard category. This report form is available from the EPA, but many states have moved to an electronic reporting system.

A tier 2 emergency and hazardous chemical inventory report must be filed if the facility produces, uses, or stores a substance on the extremely hazardous substance list in the threshold of the planning quantity, or possesses a chemical for which a safety data sheet is required in the amount of ten thousand pounds or more. The tier 2 report form is available from the EPA, but most states have moved to an electronic reporting system.

Annual reporting under section 313 of the EPCRA involves the toxic release inventory program. A facility is covered if it has ten or more employees or the equivalent of twenty thousand hours per year, if the primary activity is described by one of the Section 313 listed NAICS codes, and if manufactures, imports, processes, or otherwise use a Section 313 chemical greater than its threshold quantity. The list of covered industries is accessible on the TRI program homepage through the EPA. A list of Section 313 covered chemicals is issued each year and can also be found on the TRI program's homepage. These chemicals cause cancer or chronic human health effects, significant adverse acute human health effects, and significant adverse environment effects (EPA n.d.). This program covers all inventory to include normal, permitted release or accidental releases. It is important to know that if an accidental release occurs, your facility will likely have reporting requirements above typical emergency numbers.

This is why emergency response planning crosses over so many other topics involved in safety management. Regardless of whether the organization has a separate environmental management department, it becomes necessary to conduct an inventory of all hazardous materials and an inventory of waste streams and aspects or environmental hazards. An emergency can facilitate an accidental release; therefore, knowing what aspects are present is vital. Table 5.11 shows a suggested aspect inventory form.

TABLE 5.11   *Aspect Inventory Form*

ASPECT	SOURCE	ENVIRONMENTAL HEALTH	SEVERITY	CAUSE EFFECT
COMMON NAME	Process	Air	Environmental impact	Activity
CAS:	Activity	Water		Failure
UN:	Inputs	Soil		Impact:
DOT CLASS:	Byproducts	Wastes	Significant impact	Negligible: ☐
DIRECT: ☐	Wastes	Noise		Moderate: ☐
INDIRECT: ☐		Odor		Significant: ☐
				Catastrophic: ☐

CAMEO is a database for managing chemical inventory and planning emergency response that is utilized by LEPCs, SEPCs, and local first responders. It was developed jointly by the EPA and the National Oceanic and Atmospheric Administration. It has four components: The first is a database of eight modules that helps with data management for EPCRA requirements; the second is a database with response information for chemicals and a reactivity prediction tool; the third component, called MARPLOT, is a mapping tool for planning and for determining potential impacts; and the fourth component, ALOHA, allows users to estimate downwind dispersion of a chemical cloud. It works with MARPLOT to display threat zones (EPA n.d.).

Typically, states have suggested plan templates that should be completed by the facility. These templates may not be specific enough for the organization's emergency response plan but should be an element completed to help local and state agencies integrate what information they need.

These plans typically call for a response point to be predetermined with a staging area. These should be described in writing and be located on a map. In addition to describing the areas with address and description, coordinates can aid in pinpointing responders who are not familiar with the facility by integrating the use of global positioning system devices with coordinates.

These templates will also ask for maps of the facility with chemical storage information tied to the map. Any chemicals that are on an extremely hazardous list or require listing on tier 2 reports should be culled for storage, location, and transportation methods to and from the facility identified on specific maps.

A large aspect of planning includes mapping an evacuation radius map of the community divided into quadrants. Each quadrant should include all known addresses, name or type of facility including residences, and contact information. Special facilities such as schools, sporting arenas, malls, or hospitals should be highlighted.

Depending on the worst-case scenario of mixed chemicals, or of a single chemical being released, special instructions for protective actions should be included. This information is obtained from the Green Book, or from the information contained in the CAMEO database.

On-hand emergency equipment and capabilities should be outlined along with the general capabilities of local agencies to include medical treatment facilities. While this information may not be shared specifically with the organization, the safety manager can confirm with local agencies and facilities that they can deal with the chemicals on hand for decontamination purposes.

Spill containment/clean-up and release protocols should be written out. Preplanning the level of incidental spill versus spill is important. Determining hazards that go along with these duties, along with protection controls, should be covered in a job hazard analysis procedure and also kept as an appendix to this plan.

Lastly, important contact information for responders and emergency agencies is also included. The organization will also need the contact information for any response partners contracted to supply equipment or personnel for either containment or clean-up. More advanced organizations also include audit results of their contracted partners to show due diligence in assuring adequate equipment and personnel resources. These may be records associated with the emergency response plan. For example, if a facility was a "substantial harm" facility under EPA regulations in the Clean Water Act, specific plans for small, medium, and worst-case scenario spills would have to include a minimum equipment and supply list. This can involve relying on local contractors or specialty companies.

STORM WATER, SPILL PREVENTION, AND FACILITY RESPONSE PLANS

Some organizations deal with storm water protection plans, spill prevention control and countermeasures, and facility response plans separately, but all must be considered in the overall emergency response plan. If the organization has separate environmental and safety departments, they must work together closely to coordinate emergency response planning and any specific plans required by environmental regulation.

All facilities must plan for the protection of storm water. This begins by identifying any aspect that could pollute drainage and controls to prevent its release from company property in storm water sewers, normal drainage into streams or intermittent streams, or leaching into underground water. These aspects will be mapped out on a storm water protection map.

Storm water protection maps show aspect locations in relation to storm sewers and drainage. They also include inflows and outflows to the property. Inflows are areas where runoff from other property enter company property. Outflows are points where runoff leaves company property. Collection points of drainage and square footage of various areas of concern are calculated. Collection points such as drainage ponds and special controls such as oil/water separators are also identified on the maps. These maps also identify the sampling points of where periodic samples will be taken.

The facility's storm water permit will require a schedule for periodic samples to be taken and examined by a lab for typical contaminants such as suspended solids or total grease, but also for specific contaminants that result from the activities of the facility. The best practice is to construct a large collection pond with a liner beneath the soil and a controlled discharge gate. In case of an emergency, one of the special positions should be to close the discharge gate of the pond so that any runoff from the emergency can be collected and release from the property can be prevented. Furthermore, storm sewer inflows would have to be protected from possible runoff. This is one example of how environmental prevention, containment, and response plans integrate with emergency response.

If a facility has an aboveground oil storage capacity greater than 1,320 gallons or a completely buried storage capacity greater than 42,000 gallons and there is a reasonable expectation of an oil discharge into or on navigable waters, it must have a spill-prevention control and countermeasure (SPCC) plan. An SPCC plan has the following elements:

- Facility diagram and description of facility
- Oil discharge predictions
- Secondary containment structures and diversionary structures
- Facility drainage
- Site security
- Facility inspections and schedule
- Requirements for bulk storage containers (example include overfill protection, leak detection, prevention measures, integrity testing)
- Transfer procedures and equipment

- Requirements for oil-filled operational equipment
- Loading/unloading requirements
- Fracture evaluations
- Personnel training
- Recordkeeping requirements
- Five-year plan review
- Management approval
- Plan certification

It is easy to see that emergency response and prevention planning will include much of the information included in this plan. The plan also sets a template for protection planning for storage of other liquid chemicals that are not oil.

A facility response plan (FRP) is required for all substantial harm facilities. A substantial harm facility is any facility that transfers oil over water to or from vessels and has a total oil storage capacity greater than or equal to forty-two thousand gallons or has a total oil storage capacity greater than or equal to one million gallons, with a lack of secondary containment, proximity to fish and wildlife sensitive environments, proximity to public drinking water, or a reportable discharge with the last five years. In addition, the EPA Regional Administrator may designate your facility as substantial harm. The FRP has specific elements that are also included in an emergency response plan. FRP elements include the following:

- Emergency response action plan
- Facility information (examples include name, owner, location, operator information)
- Emergency notifications
- Equipment lists
- Personnel and responsibilities
- Evidence of equipment and personnel availability
- Evacuation information
- Identification and evaluation of potential discharges
- History of previous discharges
- Identification of small, medium, and worst-case discharges and actions
- Discharge detection procedures and equipment
- Plans for containment and disposal
- Inspection procedures and schedule
- Training plans and logs of activity such as drills and exercises
- Mapping to include facility, surrounding area, evacuation routes, and drainage
- Security measures

These elements also impact the overall emergency response plan, and regardless of a requirement to complete an FRP, elements such as procedures for spills of varying levels and security measures for critical infrastructure at the facility are important. These elements are also required for hazardous waste storage areas or areas for recycled items that would otherwise be hazardous.

EMERGENCY RESPONSE PLAN PROGRAM EFFECTIVENESS

Program metrics for emergency response still follow the same organizational structure as all programs. Metrics are categorized as management duty criteria, operational and cultural criteria, and status or performance criteria. After-action reviews and periodic program effectiveness audits are the two methods for enacting continuous improvement and management review of the plan.

An after-action review is an analysis of occurrences that look for improvements in planning, training, and operations based on observations of actual emergency response, drills, and exercises. When possible, someone is designated to be the observer and record and analyze event operations. Even when this occurs, post-event meetings can include personnel involved and can be conducted as conceptual stage meetings for analysis and progress into actual analysis input. The goal is to cull hazards, mental and physical demands, and desired and undesired behaviors and to plan for abatement of hazards and undesired behaviors. Proper analysis can reveal equipment and training needs, develop procedures, and develop drills and exercises that can counter mental and physical demands and undesired behaviors. Observers can utilize an observation form such as the basic job hazard analysis form. Table 5.12 demonstrates such a form.

TABLE 5.12 *Job Hazard Analysis Form*

TASK	CRITICAL STEPS	HAZARDS	DEMAND RATING	BEHAVIORS		COUNTERS
				DESIRED	UNDESIRED	
			Physical: Mental:			

Demand can give a training instructor the justification to increase physical and mental stress during practical training. Along with identifying the desired and undesired behaviors associated with the main tasks, more realistic training scenarios may be developed. For example, when another person closely known to the rescuers is injured, the mental demand or stress increases tremendously and can facilitate undesired behaviors. Correlating these in analysis can suggest training scenarios where expected operations are interrupted by similar unplanned events.

The program effectiveness audit for emergency response planning can center on the measurement of the criteria arranged in Figure 5.2.

Management Criteria

No.	Criteria	Score (1-5)	Notes
1.	Are goals for drills, exercises established?		
2.	Are goals met?		
3.	Percentage of training completed to all levels of personnel: _____ Percent		
4.	Percentage of rescue procedures developed		
5.	Equipment needs are met		
6.	Plan reviewed by external agencies		
7.	Representation at all LECP meetings Participation: meets or exceeds		
8.	After action reviews conducted and recorded to all events, drills, exercises		
9.	Changes to plan reviewed and based upon after action reviews		

Operational/Cultural Criteria

No.	Criteria	Score (1-5)	Notes
1.	Authorized personnel demonstrate assigned duties		
2.	Competent personnel demonstrate and understand assigned duties		
3.	Evacuation procedures carried out adequately AVG Evac time:		
4.	Shelter in place procedures carried out adequately		
5.	Ranking of importance of emergency response by personnel: AVG score		
6.	Ranking of adequacy of emergency response plan by personnel: AVG score		
7.	Ranking of emergency response equipment adequacy by personnel: AVG score		

No.	Criteria	Score (1-5)	Notes
	Status Criteria		
1.	Number of incidents requiring special rescue procedure usage: Actual:		
2.	Number of emergencies utilizing the emergency response plan: Actual:		
3.	Number of drills/exercises conducted in year: Actual:		
4.	Occurrence Breakdown: Natural Disasters: Types: Threats: Types: Unintended Consequences: Type:		
5.	Actual number of impairments:		
6.	Actual number of improper impairment activations:		
7.	Actual number of times an outside agency responded to facility:		
8.	Actual number of false alarms:		

FIGURE 5.2   Emergency Response Plan Audit

The program effectiveness audit provides a clear picture of the potential for the emergency response program's success. This program predicts, prevents, plans, and prepares for emergencies. Preparing for unintended consequences based on equipment and activity also helps in preventing incidents because the scenario preparation identifies shortcomings in the controls of hazards. After-action reviews are a must for continuous improvement in emergency response planning, but they also, in turn, drive assessment for safety dilemmas in general. This is why the emergency response plan is considered the most important core management program for any company or facility.

1. From an administrator viewpoint, what do you think the most important activity will be for you to ensure is occurring and why?
2. Is the emergency response plan the most important core management program for the safety department to be concerned about? Justify your answer.
3. Do you think the occurrence of drills and exercises is an indication of management commitment to overall safety of a facility? Why or why not?

CONCLUSION

Emergency response is a critical duty for the safety professional. Proper emergency response planning, preparation, and response truly earns trust with all associates and therefore positively impacts safety leadership. It is a complex duty that requires a broader knowledge base of other legal requirements such as state laws, local ordinances, and environmental protection standards in addition to OSHA regulations. Partnering with local responders and emergency planning commissions is also a necessary component of a proper plan. Safety professionals are truly first responders in private and public settings.

REFERENCES

Alkov, R. A. 1972. The life changing unit and accident behavior. Naval Lifeline, September–October.

Conroy, Mark, T. 2008. Fire extinguisher use and maintenance. In *Fire Protection Handbook*, 20th ed., edited by Arthur Cote, 1771–1792. Quincy, MA: National Fire Protection Association.

Environmental Protection Agency. n.d. "What Is the CAMEO Software Suite?." https://www.epa.gov/cameo/what-cameo-software-suite#other

New York Office of Mental Health. 2018. "Violence Prevention: Risk Factors." https://www.omh.ny.gov/omhweb/sv/risk.htm

National Fire Protection Agency. 2012. NFPA 101: *Life Safety Code*. Quincy, MA: Author.

Occupational Safety and Health Administration. *Citations; Notices of de Minimis Violations: Policy Regarding Employee Rescue Activities* (29 CFR 1903.14). Washington DC: US Department of Labor.

Occupational Safety and Health Administration. *Exit Routes and Emergency: Planning Coverage and Definitions* (29 CFR 1910.34). Washington DC: US Department of Labor.

Occupational Safety and Health Administration. *Design and Construction Requirements for Exit Routes* (29 CFR 1910.36). Washington DC: US Department of Labor.

Occupational Safety and Health Administration. *Maintenance, Safeguards, and Operational Features for Exit Routes* (29 CFR 1910.37). Washington DC: US Department of Labor.

Occupational Safety and Health Administration. *Emergency Action Plans* (29 CFR 1910.38). Washington DC: US Department of Labor.

Occupational Safety and Health Administration. *Fire Prevention Plans* (29 CFR 1910.39). Washington DC: US Department of Labor.

Occupational Safety and Health Administration. *Medical Services and First Aid* (29 CFR 1910.151). Washington DC: US Department of Labor.

Occupational Safety and Health Administration. *Portable Fire Extinguishers* (29 CFR 1910.157). Washington DC: US Department of Labor.

Occupational Safety and Health Administration. *Bloodborne pathogens* (29 CFR 1910.1030). Washington DC: US Department of Labor.

Occupational Safety and Health Administration. 2007. *Letter of Interpretation: Brogan Letter.* Washington DC: US Department of Labor.

Sprague, Jeffrey, R. and Hill M. Walker. 2005. *Safe and Healthy Schools: Practical Prevention Strategies.* New York: Guilford Press.

Centers for Disease Control and Prevention. 1996. "Violence in the Workplace: Risk Factors and Prevention Strategies." *National Institute for Occupational Safety and Health* 96–100. https://www.cdc.gov/niosh/docs/96-100/default.html

US Department of Homeland Security. 2012. *Best Practices for Mail Screening and Handling Processes: A Guide for the Public and Private Sectors.* Washington, DC: Author. https://www.dhs.gov/sites/default/files/publications/isc-mail-handling-screening-nonfouo-sept-2012-508.pdf

Von Oppenfeld, Rolf, & Testlaw Practice Group. 2011. "Emergency Planning and Community Right-to-Know Act." In *Environmental Law Handbook*, 21st ed., edited by Thomas F. P. Sullivan, 893-934. Lanham, MD: Bernham Press.

CHAPTER SIX

Structuring the Investigations Initiative

FOREWORD

Most of the duties a safety professional conducts relies on investigation skills. Just as a new safety manager must conduct a hazard inventory to assess exactly what he or she must combat to make the facility safe, so begins the investigative nature of safety management. Just as initial assessment begins with hazard recognition duty, investigative skills begin with hazard recognition.

Investigations must be accurate and credible to produce effective counter-measures. Safety investigations are different from quality and most engineering problems in that they involve humans and not measures of specification.

This chapter will cover the basic considerations and activities required to manage incident investigations and proactive audits to achieve accuracy and credibility and to produce thorough and effective countermeasures.

Objectives
By the end of this chapter learners will be able to do the following:

1. Define investigations
2. Identify the levels of incidents
3. Formulate avenues of reporting
4. Assign responsibilities according to competency
5. Design credible methodology for accurate investigations
6. Utilize incident mapping for accuracy and continuous improvement
7. Formulate proactive investigations
8. Recognize the goals of auditing
9. Develop an audit system

LEARNING PLAN

LEVEL	ULTIMATE OUTCOME	CLAIM	LEARNING TYPE	ASSESSMENT
F M	Define investigations Identify the levels of incidents Formulate avenues of reporting Assign responsibilities according to competency	Safety managers are investigators and must be capable of identifying an investigation, investigational levels, and formulating avenues of employee reporting.	AC/CT	The learner will formulate a matrix for assigning the level of investigation to criticality of the incident and formulate avenues of employee reporting and associate duties based on competency levels.
U	Design credible methodology for accurate investigations Utilize incident mapping for accuracy and continuous improvement	Results of investigations varying between situations and investigators must provide credible results and utilize tools for accuracy assessment and continuous improvement.	AC/CT	Learners will develop a causal analysis section of an investigation report to include incident mapping. Learners will analyze a given scenario for causal analysis and mapping utilizing their developed forms for the task.
U	Formulate proactive investigations Recognize the goals of auditing Develop an audit system	Safety professionals utilize auditing as a form of proactive investigation and review for continuous improvement.	AC/CT	Learners will develop a compliance level and a program effectiveness audit instrument for a given example.

Learning Plan Legend

Level:

F: Foundational outcomes: Basic abilities

M: Mediating outcomes: Progress through a developmental model; interpret, analyze, evaluate progressively challenging claims and arguments

U: Ultimate outcome: Navigate most advanced arguments/claims

Type of Learning:

CR: Critical reading: The ability to read, process, and understand the meaning of written information

IL: Information literacy: Locating and selecting suitable information for a task; evaluating appropriateness/validity of information sources

AC: <u>Application of concepts</u>: Ability to apply discipline-specific knowledge/skill to tasks/situations important to the discipline

CT: <u>Critical thinking</u>: Ability to apply a concept to a vague or argumentative claim without a creative leap

AT: <u>Analytical thinking</u>: Ability to critique/analyze situations using a concept or model

CA: <u>Creative application</u>: Ability to apply a model/concept in a new way/to an unrelated situation or scenario; involves creative leaps.

INVESTIGATIONS

An investigation can be reactive or proactive. Proactive investigations are typically audits or a review for a set of written or formalized criteria. An incident investigation is reactive to a known incident. It is an identification of chronological observable events, aligned with all conditions and factors present per event that produces a countermeasure from a causal analysis. Once a countermeasure is implemented, the counter must be assessed or audited for effectiveness and improved on, if possible.

The investigations program will then be a two-pronged program consisting of incident investigations and proactive auditing. But the goal is for the organization to learn from its experiences to become more efficient in its activity. To accomplish its goal, the system must gather, document, and track the relevant data; investigate certain occurrences; implement countermeasures, improve the counters; and communicate the learned data in a way that standardizes its successes.

Because of management being the activity of allocating resources to the right problems efficiently, investigations must be accurate and credible in the production of countermeasures. Otherwise, successful safety management is not possible. If counter after counter falter and do not produce positive loss prevention, support for future counters becomes increasingly difficult to obtain. The investigations program must ensure it can detect occurrences, document and track the needed data to analyze it, accurately produce investigations that have complete and effective countermeasures, and improve and communicate the learned information to its members. This is accomplished with development of investigational skills, establishing effective methodology for credible analysis, and review of results for continual improvement.

Investigations and investigational skills appear in safety management in most of the safety department's activities. Programs such as behavior-based safety, hazard recognition, human factor assessments, training needs evaluations, and walk-through inspections, are all good examples of investigations other than incident investigations.

Objective investigations that produce effective countermeasures increase the safety department's overall level of influence. As safety produces more and more positive results, associates begin to accept safety's message more and more. It propels the power of influence to the performance level. The more safety has positively helped individual associates, the more the level of influence increases to the personal level. This makes the investigations program vital for safety success.

One key element to program success is the development of basic investigational skills in personnel who will investigate and participate on any countermeasure plan or problem solving committee. Investigational skills begin with the skills outlined in Table 6.1.

TABLE 6.1 *Investigational Skills*

HAZARD, THREAT, VULNERABILITY RECOGNITION	The ability to define, recognize, and classify hazards, threats, and vulnerabilities
CAUSAL RECOGNITION	The ability to make connections between factors and conditions that contribute to cause
ANALYTIC DEDUCTION	Making a causal connection between observed factors and conditions in a predictive manner
ANALYTIC REDUCTION	The process of breaking an incident down into incremental steps and analyzing each step involving mapping and using a standard causal analysis methodology. This involves deductive and inductive reasoning
CAUSAL ANALYSIS	Examining events for causation
ABATEMENT STRATEGY	Developing countermeasures for problems identified through causal analysis to hazard, threats, and vulnerabilities
HANDLING INVOLVED PERSONS	Identifying, classifying, approaching, and conversing with involved subjects in a manner that facilitates the gathering of objective facts
WRITING SKILLS	The ability to communicate events, causation, facts, and solutions in narrative form
GROUP LEADERSHIP	The ability to work with cross-functional groups of personnel
REVIEW/PROGRAM MANAGEMENT SKILLS	The ability to review investigational initiatives for continual improvement

REFLECTION 6.1

1. If you could only classify investigations into two types, what types would they be and why?

2. Why would objective investigations earn respect and trust with associates? How would this aspect of a proper investigations program impact safety leadership?

GATHERING INFORMATION

The investigations program begins with obtaining reports of information as early as possible in the investigational timeline. The sooner the information can be gathered, the more preventive the countermeasures will be. Figure 6.1 shows the investigational timeline.

The key to obtaining reports made by associates is to encourage participation by overcoming the barriers that discourage associate reporting efforts and to systemize the uncovering of problems. It begins even at the first stage of the investigational timeline. Attitudes influence behaviors. Therefore, the goal of safety leadership, which is to make safety a virtue that does not yield out of the decision-making circle, plays a vital role in associate reporting and participation levels.

We encourage associate reporting by creating avenues for approach that appeal to all the various preferences that we know of. The premise is that it is better to offer an avenue of reporting that may be used rather than to overlook one report. Therefore, we will not rely on an open-door policy, no matter how well we perceive our own approachability. To understand this more in depth, we must examine the reasons associates do not report incidents to safety.

People do not report incidents or voice concerns, ask questions, or make complaints due to fears and personal preferences. Fear of reputation can include many issues such as the following:

- Fear of being known as a klutz
- Fear of being known as a company person
- Fear of being a tattle-tale
- Fear of being known as a trouble maker
- Fear of medical personnel
- Fear of ruining a record
- Fear of paperwork or feeling of inadequacy in filling out reports (Bird, Germain, and Clark 2003)

Personal preferences may include preference in avenues of reporting, persons approached for reporting, concern for anonymity, and concern for

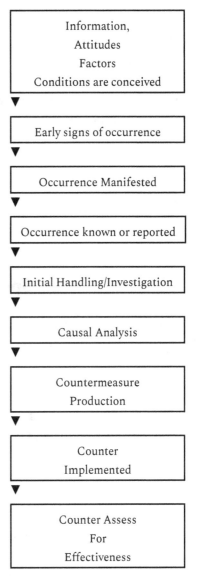

FIGURE 6.1 Investigational Timeline

documentation. Safety must overcome the fears and preference issues to receive as much information as possible.

First, it must establish an open-door policy that encourages and establishes the right to work safe. The basic elements of an open-door policy cover who reports what information to whom and when it is appropriate to do so. It also establishes proper use of the chain of command in regard to obtaining satisfactory answers. But a must-have element is a guarantee of no retaliation for questioning or reporting concerns, hazards, occurrences, or any other issue that the associate may want to bring up. Our desire is to invite associates to interact freely with safety professionals and all management. But we also have a secondary desire to establish a duty to participate in this manner and inform about the right to work safe.

With the open-door policy, as with any reporting avenue, we must overcome the barriers to reporting and participation. If we are writing policies for the associates to reference rather than establishing a documented rule for governing the company, the open-door policy must be easy to read. Most people do not want to read writing in paragraph format. It plays to the fear of paperwork, or a feeling of inadequacy in reading comprehension. We can utilize a technical advertising style of writing that culls key points and brings attention to certain items. We can bold some words, use colors, provide information in lists and bullet-point formats, or provide charts and graphs that further explain the written text. Figure 6.2 represents one possible format for an open-door policy.

Speak Up! Work Safe!

XYZ Corporation: Open Door Policy

Work with Supervision

Your **open and honest participation** is the key to creating **a safe and rewarding work environment**! All associates are empowered to **report concerns, questions, hazards, injuries, or make suggestions for improvement by meeting with any member of supervision or safety management at any time**. Associates should address safety related issues with a frontline supervisor and/or safety officer first. *If the response is not satisfactory, the associate may take the issue up the chain of authority without concern for retaliation.*

Any XYZ Associate that feels they have been retaliated against should report this to the Human Resources Director. The Director may be contacted privately by utilizing the following phone number (555-304-1584) or by email at: human-resourcesdirector@xyzcorporation.com

Empowered for Action

All Associates at XYZ have the right to work safe! When the situation warrants the **associate may take immediate action to ensure their own safety and the safety of others or to STOP WORK and immediately have the safety department assess the situation and make redress.**

Official Response

The safety department will receive your input professionally and communicate an official response as soon as possible. Team work begins with open and honest dialogue. Team work will ensure that XYZ Corporation can address issues in our workplace by finding solutions that work!

You can report anonymously to the safety manager through email or a private call.

safetymanager@xyzcorporation.com or (555) 000-0123.

Policy Outcomes
XYZ Corporation work environment descriptors:

- Associates are open and honest in communication with management
- Associates are comfortable stopping an operation for safety concerns
- Workforce Associates and Managers hold each other to the highest level of conduct

Advice and positive criticism is always accepted without the fear of retaliation

FIGURE 6.2   Open Door Policy

The goal of establishing the reporting system is two-fold: to gather information to find out what is happening to respond appropriately and to predict future issues and establish trust, which in turn increases the ability to gather the relevant information. Without a reporting system that encourages multiple preferences to participate, safety may miss vital information that can lead to countermeasures that prevent loss. An open-door policy is not enough.

An open-door policy does not overcome enough of the reporting barriers to be a stand-alone system. Workforce associates may simply not like the safety manager. They may also still fear reputation issues. If they are seen communicating with company managers, they may be labeled a rat, or be viewed as "sucking up" to get a promotion. Even if this is done in jest by their fellow workers, it is, often enough, too much of a discouragement.

Anonymity is the most important piece to earning enough trust for reporting to increase. Therefore, avenues that do not identify the person participating can be important. With the open-door policy, it is beneficial for associates to be able to contact management in a more private manner, such as through a cell phone.

Some may want to make their report on an official form. It may be that they feel it is more likely to be responded to, or that a written report shows documentation, or makes it more official or important. But this avenue of reporting a concern, asking a question, or making a complaint is important in earning trust. If employees cannot make complaints or voice concerns, how can they truly be empowered? Safety managers desire to have employees bring these issues to them first so that they may make all attempts to counter the problem once it is confirmed. If associates cannot bring problems forward, then they will take the problems outside the organization to be fixed, such as making the report to OSHA.

The safety concern card must still overcome as many barriers as possible, such as being short to overcome people's fear of red tape, making as many items as possible "check-box items" to simplify the form, and not a requiring name to protect anonymity if desired. Figure 6.3 is one possible safety concern card.

Hazard Report Card

Location/Production Line: _____ Authorized Associate: _____

Date: ___/___/___ Competent Safety Associate: _____

Hazard Category	Source	Control	Status
☐ Compression	☐ Radiation	☐ Corrected	
☐ Penetration	☐ Walking/Working Surface	☐ Action	
☐ Impact	☐ Electricity		
☐ Chemical	☐ Animal, Insect, Vermin		
☐ Respiratory	☐ Biological		
☐ Temperature	☐ Noise		
☐ Visibility	☐ Physical Ergonomic		
☐ Biological	☐ Physiological Ergonomic		

SAFETY DEPARTMENT/Competent Safety Associate USE ONLY

Hazard Found By Associate   IC #_____   Follow-up Date:___/___/___

Yes ☐   No ☐

Hazard Timely Reported   Hazard Level   ☐ 1   ☐ 2   ☐ 3

Yes ☐   No ☐

Problem Solving Committee Assigned

Yes ☐   No ☐

Countermeasure Plan Completed

Yes ☐   No ☐

(Dotson, Blair, Rawlins,& Rockwell, 2018).

FIGURE 6.3   Hazard Report Card

It is typically said that only the disgruntled and the extremely satisfied associate will utilize the safety concern card. But it is still worth having in place since its use or non-use in comparison to other avenues of reporting and measures of participation can give insight toward morale about safety and participation. Regardless, if a disgruntled associate did not have an avenue to vent or voice concern, where would he or she go to vent? Would the situation continue to fester without a chance to redress the situation? So, the safety concern card is there to fill a void for those who will not utilize the open-door approach but want an official avenue to communicate.

As a continuation of the open-door policy, but to fill the void for those associates who are not comfortable coming to supervision, we can utilize an active associate safety committee for another avenue of reporting. Safety committee representatives can also receive questions, concerns, or complaints as part of their duty to be a liaison between associates and safety management. This is called a peer reporting initiative. Many associates may be more comfortable going to a peer rather than any other method of reporting.

So far in the Employee Oriented Safety system, we have formal avenues for reporting and correcting level 1 hazards and avenues for communicating with the safety department about questions, concerns, or complaints. In our attempt to establish trust and address any concern in regard to safety, we have redundant avenues for reporting that attempt to cover all preferences and overcome barriers. In regard to incidents or observable occurrences that have potential for negative result, such as injuries, near misses, property damage events, security breaches, or threats and other workplace violence situations, or any emergency where the importance level is so high it warrants an immediate and verbal report to supervision, these must be covered in a policy statement that goes along with associates' duties to participate in any investigation as requested in an open and honest manner.

It then becomes the duty of the immediate supervisor to notify the safety department and begin the investigation by handling the immediate scene and filing the first report of incident to the safety department.

The immediate supervisor is a secondary investigator. He or she has more intimate knowledge of the people and workstations than the safety department and has a duty to be involved in occurrences in his or her jurisdiction; it also increases his or her buy-in to workplace safety in general. Their immediate role is to do the following:

1. Take control of the scene and assign others to aid appropriately and to immediately put into action the emergency response plan. The priority is to always begin summoning more advanced care if needed.
2. Ensure safety of injured and responding personnel. Making the scene safe for responders and, if necessary, remove or direct immediate rescue of injured persons so that first aid may be delivered is crucial. Life always takes precedence.
3. Preserve the scene and protect evidence; this is secondary to life. In this action the person handling the scene should control access to the scene, identify witnesses and separate them as soon as possible, and begin scene documentation. Documenting the scene means taking pictures, making sketches, taking notes, and marking key areas or evidence.
4. Update and turn over scene control. When the lead investigator arrives, the supervisor updates the lead investigator and begins to complete these duties or aids the lead investigator as directed. This may include helping to interview witnesses, gather evidence, perform audits, file reports, or fulfill any number of possible needs.
5. The supervisor files a first report of incident to the safety department.

The first report of incident to the safety department is a bare bones basic set of facts that the lead investigator will need to follow up on the incident, file initial reports or notify an insurance carrier, and begin incident documentation—enough to require the supervisor to fulfill his or her obligation to supervised personnel and gain buy-in to be concerned about incidents. The data required, at a minimum, is based on a scenario of a minor incident that can be handled in the absence of a safety department representative and notification can be made at a later time of that day. Critical incidents that are more than first aid or a minor loss should require both immediate reporting to safety and a response to the scene by a safety representative, even if safety is not present in the facility. If a safety representative is present, he or she should respond to all incidents.

The first report to safety should include basic information about the incident, such as time and date of occurrence, how it was reported, who was involved, identification of all involved persons, a basic set of facts or description, identification of any outside agencies responding, identification of any injured persons, level of severity, level of treatment, and any information when injured persons are removed from the scene. Figure 6.4 represents one possible First Report of Incident (FRI) to safety.

FIRST REPORT OF INCIDENT   IC#

Type of Report:(circle one)   **Accident w/Injury   Near Miss   Complaint   Security**

Date of Incident: ___/___/___ Time of Incident: _____ a.m./p.m.

Injury: Y or N   Property damage? Y or N   Pictures? Y or N

Person filing report: _____

EMPLOYEE INFORMATION: Shift: _____

Full Name: _____   Phone: _____

Address: _____   Dept: _____

_____ (circle one) Full Time or Part Time

D.O.B. ___/___/___   SSN# _____

Date of Hire: ___/___/___   Circle One: Male or Female

First Aid at scene: yes no   First Aid provider: _____

Injured removed to: _____

Injured or Deceased removed by: _____

Injured Employee: _____ S.S.N# _____

Address: _____

D.O.B. ___/___/___ Date of Hire ___/___/___ No. of Dependents _____

Emergency contact notified: Y N   Substance Test: Y N Pay Rate: _____

.

Describe Injury: Body Part: Appeared Severity:

Details of Incident: Include actions of employee just prior to injury and the PPE involved. List all property   damaged. Describe location with machine number/ area.

_____

_____

_____ P P E :

Property Damaged:

Involved Persons: (name, ID # if applicable, address, phone #, d.o.b. for non-employee witnesses, attach statements to report form)

Place Involvement Code before Name:
W for witness   **I for involved**   **P Person of Focus**

1. _____

2. _____

3. _____

4. _____

Attach: (circle attached)   JHA   BBS Review   Written Procedures

S.O. _____   **First Report of Incident**   Page 2

Draw accident: if motor vehicles or fork-trucks are involved. Indicate North by an arrow. Not to scale.

<u>Safety Department Use Only:</u> Full Investigation Ordered: ☐ Yes ☐ No

(if NO, complete following, attach causal analysis, countermeasures report)

WORKSTATION: NAME _____

Environmental Conditions:(circle) Indoor or Outdoor Temp. ___ WBGT. ___ Wind Speed

Noise Rating: (circle) High (> 95 DBA uncorrected) Medium (> 80 DBA uncorrected) Low (< 80 DBA uncorrected)

Physical Demand Rating: (circle)   High () Medium () Low ()

Cognitive Demand Rating: (circle) High () Medium () Low ()

Behavior Based Composite Score: _____

General Description: (circle) rain snow ice icy wet dry cloudy sunny windy fog dusty

Lighting: (circle) dawn daylight dusk   dark- outside lighting on   indoor plant lighting on

dark- outside lighting off   indoor plant lighting off

dark- outside lighting inop   indoor plant lighting inop

Risk Assessment

Risk from Employment: Yes ☐ No ☐   Risk Personal to Injured: Yes ☐ No ☐

Employment Risk Factors:	**Personal Risk Factors:**
☐ Height ☐ Exertion ☐ Surface ☐ Motion	☐ Medical ☐ Mental
☐ Confined Area ☐ Congested Area	☐ Failed to take medication
☐ Noise ☐ Heat/Cold ☐ Weather	☐ Balance Condition ☐ Hearing ☐ Eyesight
☐ Other: _____	☐ Previous Injury
	☐ Other: _____

Probability: High☐ (3)   Medium ☐ (2)   Low☐ (1)

Experience: Routine Occurrence ☐ (3)   Occasional Occurrence ☐ (2)   Rare Occurrence ☐ (1)

Exposure: Several Persons ☐ (3)   Few Persons ☐ (2)   Single Person ☐ (1)

Risk Rating Total:

FIGURE 6.4 First Report of Incident

First reports are basic and simple to file. A first report includes just enough basic information to allow the lead investigator to pick up the case. It is up to a lead investigator to get in depth on documented information and perform causal analysis. These reports are intended to be completed on the scene and should be included in a basic investigator kit. These kits are the necessities that are required, besides first-aid and trauma supplies that the supervisor may need. For example, the kit could include items such as the first report of incident, a pen, a disposable camera or digital camera, an evidence-collection bag, emergency planning maps, roll-call sheets, copies of JHA forms or other forms, and protective gloves and other appropriate PPE such as protective coveralls, aprons, or medical face shields.

The first report of incident can include incidents that are less complex and merely include a causal analysis section and countermeasures report from the full investigation report. However, it is important to note that it can be easy to overlook collecting important data such as cost.

> ## REFLECTION 6.2
>
> 1. One criticism of developing a safety concern card system is that it might encourage complaints? Why is receiving, even complaints, important to leadership?
> 2. Developing multiple avenues of reporting based possible workforce preferences is an example of practicing inclusion. Is this transformational leadership practice and why or why not?
> 3. How would you motivate associates to fully participate in reporting safety questions, concerns, or suggestions?
> 4. Why is an open-door policy not sufficient for hazard recognition?

DOCUMENTING EXPERIENCE

Managers must create sustainable systems that survive personnel changes and adapt to diverse situations. At the foundational base for any managed system is a method for documenting and recording all occurrences or experiences and activities for the areas managed by the system. This is the incident control log (IC). This log is the first stop for the safety department to record its activity whether it is an injury incident, an incremental spill, a proactive audit for hazards, or a complaint from an associate.

This avenue of documentation allows the safety manager to examine for types and frequencies of activity, justify resource requests, track activity that needs to be completed, and helps the manager allocate resources efficiently. This log will record and track overall first-level information about occurrences. This data may include items such as date and time of occurrence, type of event, reporting method, primary involved person, the reports associated with the incident, and even outcomes. Its primary use is to assign each occurrence a unique identifier so that all subsequent activity and paperwork can be identified and tied to that particular occurrence. This log can be as condensed and complicated as the coding methods for recording the data that will be documented. Figure 6.5 reflects on possible example of an IC log.

IC Number ????	Date/Day	Type Incident	Involved Perso	Superviso	Injury/: (Y/N)	Investigatio (Y or N)	Dept	Outside Agenc	Open/Close	Notes:
??0001	1/1/Tue	Alarm/Entry	N/A	N/A	N	N	Office	PD	Closed/Fals	Pres. Schumann entered
??0002	1/1/Tue	Alarm/Power	N/A	N/A	N	N	All	AEP	Closed/Fals	Power outage
??003	1/2/Wed	Injury	Randall, Dave	.Dixon	Y/Y	Y	Weld		Closed	
??004	1/3/Thur	Prop Loss	Fee, Steven	Sears	N	Y	Ship	Eagle Trucking	Open	Forklift into trailer
??005	1/5/Sat	Near Miss	Chadwell, Shar	Sears	N	Y	Ship	N/A	Open	Forklift rolled out bay door
??006	1/6/Sun	Injury/Burn	Thomas,Scott	Simms	Y/Y	Y	Weld	N/A	Closed	
??007	1/7/Mon	Near Miss	Welker,Lisa	Ford	N	N	Ship	Eagle Trucking	Closed	Forklift almost hit by truck
??008	1/7/Mon	Near Miss	Moore, Ron	Dixon	N	Y	Weld	N/A	Closed	machine malfunction
??009	1/7/Mon	Theft	N/A	Gene	N	Y	Maint	N/A	Open	tool theft
??010	1/8/Tues	Audit	N/A	N/A	N	N	Weld		Open	machine guarding

FIGURE 6.5   IC log

Source: Dotson et al. 2018.

The IC log becomes the first step in response or in assigning or issuing out an order for activity from safety. As soon as an occurrence is known, it becomes logged with an IC number.

> **REFLECTION 6.3**
>
> 1.   Documented experience is vital for strategic planning. In what ways?

INVESTIGATIONAL DYNAMICS

The real difference in an investigation and an assessment of an occurrence is that an investigation is much more complex. Causal analysis will produce countermeasures formulated from an abatement strategy, and this process can be completed from ad hoc committees or follow a standard process with cross-functional input. This is very important in managing investigations. Resources are limited. Therefore, we must designate which type or level of incidents will require full investigation reports completed and/or when cross-functional cooperation is needed to analyze for cause and formulate countermeasure recommendations. Some organizations may require all countermeasures to be reviewed by a cross-functional system safety committee when a moderate modification or major modification is being made to a machine or process. Some investigations will be complex enough to require expertise from cross functional personnel to identify causation and, in this case, would justify the same committee to formulate countermeasures. But some investigations can be handled by the lead safety investigator and make countermeasure recommendations that would only need informational review before enacting. This is the basis for defining critical incidents versus minor incidents. The lead safety investigator makes the judgement as to whether the situation presented warrants critical status.

Credibility is the major concern for an investigations program. Credibility is the propensity for different investigators to arrive at the same results when presented the same or very similar incidents. A manager is always tasked with allocating resources efficiently. This means resources are aiming at the right problems and having a positive effect. This goal of management demands that the investigations produced by an organization be completed with the same methodology and follow the same strategy with controls that ensure accuracy. In simple terms, the investigators must follow significantly similar methods of investigation and analysis so that countermeasures are valid and reliable.

If credibility is not designed into the investigations form and program, countermeasures will be inconsistent and not complete or accurate enough to increase confidence that the incident has been corrected. Morale toward safety, or the feeling that a place, task, or machine is unsafe, impacts production frequency and quality. In addition, the safety manager's and the safety department's reputation for correcting problems is lowered and upper-level support for investments needed for countermeasure implementation become increasingly more difficult to obtain as confidence in the safety department or the manager's skill is reduced. Furthermore, investigations become viewed as less objective and lose credibility in general with the workforce. Rumors and accusations of favoritism become problems and in turn reduce participation in safety efforts.

Designing credibility into an investigations program begins with strategic planning and establishing the goals of the program and of investigations; the measures that will be tracked; the model for causal analysis; the duties and responsibilities at the authorized, competent, and administrator level; and the problem-solving methodology for countermeasure formulation and after-implementation review.

The goal of the investigations program is to learn on an organizational level so that future incidents of similar facts do not occur. Learning on an organizational level means that the results of causal analysis are disseminated in a way that cross-functional structures retain enough of the lesson learned to enact the fix and sustain its requirements. Measuring the degree to which this occurs means that a method of identifying and tracking "repeat" incidents must be developed and reported on for overall success of the program.

Investigations on a case-by-case basis need to be accurate, or timely, complete, and effective. Therefore, each investigation will have to be assessed for timeliness and thoroughness and on how well the countermeasure worked. Assessing whether the investigation was completed and the countermeasure implemented in a timely manner is not complex. The safety manager can set a reasonable timeframe based on the totality of the incident and consider whether repeat incidents occurred prior to this completion. All incident investigations and audits will produce a countermeasure report. This countermeasure report will be considered closed once the counter is implemented. This allows us to assess timeliness as well as the commitment of management to safety by measuring resource allocation.

Once a countermeasure is in place, the organization must go back and assess it for effectiveness to make minor adjustments. So, how long do we wait? We can walk any modification through the system safety review process and then examine at the two-week, thirty-day, and ninety-day mark. If the countermeasure is deemed successful or modified to improve effectiveness to the desired level, the counter must become a standard of practice.

In many organizations a countermeasure fails to become a standard and, once completely implemented and repeat incidents seem remote, the counter falls to the wayside. The organization must develop specific standards for machines, processes, work practices, and even standards of design and fabrication that include the best management practices (BMPs) learned from experience. Once cross-functional teams develop these into the standards, the teams must be trained on the standard. Standards become part of the audit criteria examined for when any machine or process, work procedure, or task is examined during modification or new launch.

REFLECTION 6.4

1. From a management point of view why is credibility so critical?
2. What methods do you think increase credibility in your investigation results?

THE INVESTIGATION FORM

The investigation form is an important tool that gathers the data that shows performance between the level 1 and 2 metrics and final performance numbers. In other words, the core management programs we are relying on to be proactive in nature must be accounted for during a reactive investigation to confirm their effectiveness, a leading precursor to positive safety performance. A leading indicator is a positive precursor that prevents or reduces likelihood of a negative occurrence. Behavior-based safety programs promote leading indicators associated with behaviors for example. If we produce an overall rating for machines, processes, or tasks based on desired behaviors, we can compare performance of machines and processes or tasks scoring higher in regard to desired behaviors against those that score lower in regard to desired behaviors. In other words, we confirm that what we assume in the beginning are positive precursors are achieving the goal. Human factors programs are another example of programs that attempt to prevent by encouraging positive precursors such as less physical demand or less cognitive stress. Therefore, the investigation form turns reactive findings into proactive measures not merely by producing counters that are aimed at reducing repeat incidents, but also by using it as a management tool for identifying if proactive measures are working. If proactive measures are not as effective as anticipated, then the safety manager must assess the core program for effectiveness and correct any weaknesses.

The box that follows reflects a version of the section of the investigation form where this takes place.

WORKSTATION NAME: _____

Environmental conditions: Indoor or outdoor   Temperature: _____ WBGT: ____ Wind speed: ____

User satisfaction score:___/ 360

Safety efficiency total: ___/ 18

Behavior-based composite score: _____

Hazard recognition accuracy/level 1 hazards: _____

General description: rain snow ice icy wet dry cloudy sunny windy fog dusty

Lighting: dawn daylight dusk   dark outside, lighting on   indoor plant lighting on

dark outside, lighting off   indoor plant lighting off

dark, outside lighting INOP   indoor plant lighting INOP

Latest usability assessment attached: Yes ☐   No ☐

In this example, critical ratings from the human factors program and behavior-based program are being assessed for comparison and analysis. The investigation form is collecting data that will be documented on a log associated with each core management program. The goal of management is to allocate resources efficiently, or at the right issue, to gain the best positive impact. If this data is not collected in an investigation, organizational experience is not documented in a way that assists management in completing its central duty. As the behavior based and human factors programs are covered in this book, the rating methodology will be revealed. Similar concepts can be developed for measuring performance of any organization program.

The safety manager is responsible for identifying the necessary information that must be collected on the investigation form. The appendix to chapter 6 contains the complete investigation form example. Understanding the sections can help in organizing the necessary data. The sections of an investigational form are as follows:

1. General incident information
2. Involved persons
3. Incident documentation
4. Evidence dispositions
5. Cost analysis
6. Causal analysis
7. Incident mapping
8. Countermeasures report

The investigational report form example included in the appendix does not utilize this list as an outline. The data and sections can be arranged in a manner that flows or matches the order the data can be relayed to relevant logs. In this report form example, it is clear the general incident information section includes identifying descriptive data, but also a workstation assessment and risk assessment. The involved persons section includes injury data and identification of involved persons and agencies. Incident documentation includes any drawings of the scene and narrative detail of the incident. Evidence disposition considers photos and items taken and serves to begin preserving the chain of custody by identifying the method of storage so that a secure storage is documented.

Evidence documentation is critical in any investigation. Typically an inventory-style form suffices. But having policy of how evidence will be stored and examined in a way that preserves its original state or documents any tests or changes to the evidence is vital in preserving its acceptance or admissibility in court or by interested parties. Documenting who examines the evidence and then documenting its state and condition on return is called preserving the chain of custody. Even in a reconstruction of workplace incidents, this can be critical not only for legal proceedings, but in determining causation and formulating corrections.

The cost analysis pulls together the direct costs that can be identified or estimated during the time of the investigation. Tracking direct costs is a primary strategy for communicating safety as an investment rather than an incidental cost to production. Safety began to be thought of cost effectiveness in the early twentieth century and became a legitimate business management concern (Bird et al. 2003). Safety managers must track direct costs per machine, part produced, or activity in a manner that allows them to predict safety costs of new contractual obligations and during considerations of modifications or new processes and machines being introduced to the facility. This can make safety a consideration in any bid or estimation process and will have positive influence on the bottom line from a cost-savings performance review. By tracking costs, the safety department can offer a savings over performance metrics review that demonstrates its investment.

The causal analysis, mapping, and countermeasures sections of the investigation form are the most critical for successful safety performance through accurate investigations. These areas help ensure credible findings and thorough analysis and provide an avenue for performance assessment and continual improvement in the investigations program. Credible findings are findings that are consistent between investigators and between incidents. In other words, all investigators are considering the same model of causation and utilizing the same strategy for abatement. Even so, causes and counters can be missed. Inexperienced investigators often examine only the final event in the incident or examine only what is readily apparent. Accurate investigations are encouraged by the design of the form, the methods utilized, and from proper review of the findings. Performance assessment and continual improvement begin with the safety manager reviewing investigation reports. This relies on mapping, narrative, causal analysis, and countermeasure review.

Mapping is a process that begins by breaking down the overall incident into smaller incremental events. Mapping documents the events in chronological order. This is required to be thorough. Each event is analyzed for causes. We can even document the findings for cause along with the event. The narrative is reviewed along with the map to ensure that for each event the basic investigative questions such as who, what, where, when, how, and why are answered if they are known. But the first step in reviewing an investigation report for accuracy is to review the narrative and map for the degree to which the events and basic investigative questions match. So, an incident map is not merely a diagram that documents findings or factors considered, it is a tool that can be referenced to reexamine and confirm, deny, or add to the findings from the analysis.

Butterfly mapping refers to the mapping methodology and is called "butterfly" or "bowtie" mapping due to its arrangement. Events that occur concurrently are mapped parallel to each other, representing the idea that two events can occur simultaneously and allow or lead to another event. A simple example may be that a vehicle accelerates through a red traffic light concurrently with another vehicle entering an intersection with a green traffic light. The map may reflect this scenario as in Figure 6.6.

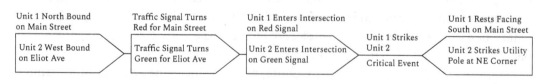

FIGURE 6.6   Butterfly Mapping

Mapping in chronological order can be very important in formulating countermeasures. Butterfly mapping also allows the investigator to consider probable events that did not actually occur in reflecting the events that occurred or could have occurred after the critical event or point of no return. This is important in the investigation of near misses. Building on the simple example of a traffic crash, the vehicles could have spun and came to rest without striking other vehicles or pedestrians. But it is very reasonable that one or both could have spilled hazardous materials such as fuel, coolant, or battery acid as well.

It is important to consider that many incidents need a sketch or other method of scene documentation, such as pictures, to supplement a map and narrative. But a map is not a sketch. A sketch might show final resting positions, progress of movements, what the area looked like, or relative positions. But a map is used for complex analysis for cause and performance review. Between each event, possible countermeasures that would prevent the event from evolving are considered. The map is now helping the investigator or team of investigators to be thorough and not overlook a critical component of the overall countermeasure plan. The overall countermeasure plan will reflect the sum of acceptable counters considered to be inserted between the events of the incident as a whole.

Causal Analysis Modeling

The safety manager is a manager of an investigations program. Successful management requires that a method of investigation and causal analysis be adopted that fits the circumstances of the organization. The methods and modeling must become standard for all those leading a safety investigation.

In order for the investigational practices to become a standard of the organization that produces credible results, training must be conducted that puts the investigators on the same page for causal analysis. Continual improvement must be practiced. Continual improvement requires the safety manager to review the accuracy of investigations after the countermeasures, or the fixes to the causes, identify weaknesses of the investigation, and produce education and training that attempts to improve investigative skills.

The Bird management model of causation is the preferred model of causal analysis. It is important to note that, regardless of causal modeling, each incremental event that comprises the progression of the incident must be examined for cause separately in order to produce a comprehensive list of all the causes present in the incident. Countermeasures to the causes can only be effective in preventing repeat incidents of like nature if they cover all of the causes present.

The Bird management model recognizes and uses most other models to further explain the levels of causation. This model correctly recognizes that incidents rarely if ever have one cause. Incidents result from an alignment of events and causal factors. This is why butterfly mapping applied to events fits with safety investigations as well. Table 6.2 reveals the application of the Bird management model.

TABLE 6.2   *Bird Management Model*

Immediate cause	Unsafe acts	Substandard conditions	
Basic causes	Job factors	Personal factors	Design factors
Root cause	Management system failures Operational Error/Organizational Error		

Sources: Bird, Germain, and Clark 2003; Heinrich, Peterson, & Roos, 1980, p.30

The Bird management model examines cause on three levels: immediate, basic, and root (Bird, Jr., Germain, & Clark, 2003). In this model root cause is not defined as the main cause of the incident or the main contributor to the final result of the incident. In the Bird management model root cause refers to the failures at the management level of the organization in regard to its duty of producing a safe and healthy workplace.

The immediate level is comprised of unsafe acts and substandard conditions (Bird, Jr., et al., 2003). Unsafe acts are those direct acts that produce the risk that results in the occurrence of an incident. Substandard conditions are those physical conditions that directly increase the risk of negative results if left uncorrected. Substandard conditions exist in direct opposition to a regulatory requirement, a company standard, safety policy, or best management practice. Usually unsafe acts are viewed from the actions of the workforce-level associate. Management and supervisory personnel can also commit an unsafe act.

Preparation for assessing unsafe acts and substandard conditions begins with reviewing legal requirements and best management practices in place in order to produce an audit or checklist style assessment for the requirements. This reveals substandard conditions while also preparing the investigator for identifying unsafe acts.

The basic level allows the immediate level to develop. It is comprised of job factors and personal factors (Bird, Jr. et al, 2003). Job factors are factual conditions that present negative risk potential akin to substandard conditions, but do not by themselves exist in opposition to a regulation or other standard. Job factors are conditions presented and necessary for production but are still required to be controlled. A good example might be the difference in operating an agricultural tractor on flat ground or on a slope.

Personal factors are the conditions or perceptions of a human motivational nature. Personal factors that exist as conditions include fatigue, stress, impairment, or frustration. Perceptions of importance and feelings of euphoria are examples of motivational attitudes that explain how unsafe acts develop or are committed. Unsafe acts can be committed with different levels of culpability. It is important to note that an organization may not be capable of controlling all personal factors. By recognizing the personal factors causal in nature to the incidents in an organization, some controls are possible.

Assessing the culpability involved with the unsafe act provides an assessment of competence. This is an important personal factor because it can be an indicator of the effectiveness to which management is establishing positive safety behaviors as a habit. It is also a guide to accountability recommendations associated with actions or inactions that are causal in nature. Identification of the culpability is used to justly fit remedial actions of training or discipline based upon the person's competency revealed in culpability. For example, a person acting with intent should be dealt with more harshly than a person who failed to recognize a potential result of their actions. Additional development of competency is warranted for the person who did not recognize the potential of the risk.

Competency is the ability to apply knowledge. Culpability is the person's mental state in regard to the results of their action or inaction. Culpability can be reckless, wanton, knowing, or intentional. Culpability is not assessed to the act itself but to the outcome or result. Applying mental state to the result is critical in determining the correct mental state and in determining remedies of a personal nature. A basic principle of justice is that the punishment must fit the crime. The same is true for accountability and disciplinary policy in private companies. When culpability is used, it is applied to the result and not the action or inaction alone.

Culpability is described in four states; reckless, wanton, knowing, and intentional conduct. Recklessness is when a person fails to perceive a substantial and unjustifiable risk that the result will occur or that the circumstances exists. Wanton activity is when a person is aware of and consciously disregards a substantial and unjustifiable risk that the result will occur or that the circumstance existed. Knowing is when a person is aware that the conduct is of a particular nature or that the circumstances exist. Intentional is when a person has a conscious objective to cause a specific result or to engage in that conduct (Kentucky Revised Statutes 502.050).

Care must be taken by the investigator to avoid application of culpability to action or inaction. The result of misapplying culpability to the action alone can be a "blaming the victim" mentality resulting in a causal determination that does not extend to the root level of the problem. Application of culpability to action is exampled when a company or investigator bases its assessment of the mere presence of the knowledge being presented in training, educational effort, or policy. Advanced safety management recognizes that this is awareness level application and must be followed up by development of values to the point that behaviors are reinforced to become habit.

An example of assessing culpability as a personal factor at the basic cause level might be considered a policy violation. If a workforce associate knew of a policy and decided to take risk that a negative result would not happen when they deviated from policy, and in fact it resulted in damaged equipment, the assessment of culpability is to the damaged equipment. Specific application of this can be exampled in regard to an employee operating a company truck in violation of the speed limit. Let's say that a minor traffic accident occurs and the driver lost control and went off the road resulting in the vehicle striking a sign post and being stuck in a ditch. The question becomes whether the operator recognized that speeding would result in a traffic crash. Obviously there is a dilemma in regard to what mental state can be proven. The operator's intent may have been to arrive on time or he may not have realized his speed. If he failed to realize his speed, then it most certainly becomes a state of recklessness. If he was intent on arriving on time or early, it still is not clear that he ignored the hazard of a wreck. He would have had to recognize, that his speed would likely result in a crash in order for it to be wanton conduct. Now there is a point that a reasonable person would have or should have known the crash was likely due to speed. In order for this example to extend to wanton conduct the investigator would have to show factual action of dangerously high speed and/or erratic control.

Root cause is simply stated as management failures (Bird, Jr. et al, 2003). The Adams model or update to the original Bird management model provides a more detailed categorical list of root causes by recognizing behaviors as operational errors and structural failures and omissions as organizational errors.

The Adams update recognizes managerial behaviors as operational errors (Heinrich, Peterson, & Roos, 1980). Operational behaviors allow basic and immediate causes to develop into an incident and originate from the duties of supervisors and managers in regard to safety leadership. These are akin to unsafe acts, but are not directly causal to the incident. Supervisors and managers may also commit an unsafe act that is direct to the incident as well. Table 6.3 shows a version of the Adams model of operational errors at the root level.

TABLE 6.3 *Operational Errors*

Operational Errors

Conduct	◦ Sets poor example ◦ Sets poor example	◦ Supervisor ◦ Manager
Responsibility	◦ Instructions not understood ◦ Instructions not accepted ◦ Pressure of immediate task obscures ◦ Out of scope of responsibility	◦ Supervisor ◦ Manager
Authority	◦ Fails to make decision ◦ Failed to delegate ◦ Out of scope of competency ◦ Out of scope of Authority	◦ Supervisor ◦ Manager

Rules/Procedures	◦ Fails to enforce ◦ Fails to follow up ◦ Uneven enforcement	◦ Supervisor ◦ Manager
Coaching	◦ Fails to explain why ◦ Failed to listen ◦ Failed to coach ◦ Inadequate instructions	◦ Supervisor ◦ Manager
Morale	◦ Tension with labor ◦ Not confident ◦ Lacks motivation to labor force	◦ Supervisor ◦ Manager
Operations	◦ Improper job placement ◦ Lack of equipment to worker ◦ Inefficient workloads/flow	◦ Supervisor ◦ Manager

(Heinrich, Peterson, & Roos, 1980, p.30).

Organizational errors at the root level appear mainly as structure issues. Table 6.4 shows organizational errors at the root level.

TABLE 6.4 *Organizational Errors*

Organizational Errors

Objectives	◦ Inadequate objectives ◦ Inadequate standards of performance ◦ Inadequate analysis/measurements
Organization	◦ Inadequate chain of command ◦ Lack of competency ◦ Over delegation ◦ Lack of delegation
Operations	◦ Improper layout ◦ Lack of policy/scope ◦ Lack of resources ◦ Procurement issues ◦ Scheduling ◦ Lack of procedural development ◦ Lack of environmental planning

(Heinrich, Peterson, & Roos, 1980, p.30).

Analyzing root causes and formulating effective countermeasures to the causes are critical components of safety leadership. The explanation is a simple but foundational principal of leadership applied to the management level of an organization; lead by example.

REFLECTION 6.5

1. What does the term "accurate investigation" mean to you?
2. How can the investigation form increase accuracy and credibility?
3. What is the difference between job factors and substandard conditions in the root–causation model shown in Table 6.2?
4. Assessing immediate causes begins with understanding what legal requirements apply to the circumstances. What strategy should you take in identifying these requirements? What management strategies might you utilize to complete the assessment of immediate causes in a timely manner?
5. How might you utilize an assessment of culpability in regard to producing counter-measures for an incident?
6. How might you use root cause analysis to assess annual performance of management level personnel in regard to safety practices?

INVESTIGATIONAL EFFECTIVENESS REVIEW

Incident investigations are a core management program and require reviews for effectiveness. We can base our review on needs found from investigations that produce less effective countermeasures and from the overall review of investigation reports. Safety managers act as reviewers of investigation reports from subordinate investigators. So, depending on the structure of the organization, the safety manager may review and make recommendation to reports submitted by subordinate safety personnel, or the lone safety professional of the facility may review his or her own countermeasures based on effectiveness and the first reports submitted by frontline supervisors.

Once an investigation has produced a countermeasure and the counter is implemented, the counter must be assessed for effectiveness, and then any modifications are made to improve performance of the counter and prevent repeats of similar occurrences. But what happens if a countermeasure is less effective than expected or desired? The review for accuracy would then refer to the map. The map would reveal where in the incident the counter was first implemented to stop or prevent the chain of occurrence. The reviewer or team would begin to assess the event for accurate facts and other possible causal factors that may have been overlooked. It may also reveal possible inefficiencies in the investigator or team's ability that needs to be improved.

Once the countermeasures are deemed effective, cross-functional team review can be a method for taking positive outcomes and turning them into company standards and completing the organizational learning process. Looking for positive precursors or the good things that the organization is doing or has done is an important aspect to investigations. In this program, assessing or discovering what makes one task or work station less dangerous than others is handled in the human factors program. However, positive precursors must be examined with causation of incidents in minds. The examination of causes can shed light on negative trends but also on what is not occurring. The trend could be behavioral, for example, or it could be a condition that presents the challenge.

Producing a causation log is necessary for tracking causation from incident investigations for easy access to causal metrics. Table 6.5 represents a possible causation log.

TABLE 6.5 *Causation Log*

IC #	MACHINE/ PROCESS	TYPE	SEVERITY	IMMEDIATE CAUSES	IMMEDIATE CAUSES	BASIC CAUSES	ROOT CAUSES
		☐ Injury ☐ Property damage ☐ Near miss ☐ Security event ☐ Environmental	☐ Light ☐ Moderate ☐ Severe	Acts	Conditions	Job factors

Personal factors | Management duty

System failure |
	Noise level: _____						
	Demands: physical: _____					Design failures	
	Mental: _____						
	BBS: _____						

We can now set an extended review date for an implemented countermeasure or multiple review dates that will look for significant similarities by location, type, severity, and cause. It is possible to assess for repeats at the thirty-day mark and then again at the 180-day mark, or action dates for this review can be set at a frequency that best serves the organization. Each critical event will be examined for effectiveness by this methodology.

Other core management programs also are tracked on the investigation report so that we may begin to assess what score or levels of score are benchmarks for performance. Figure 6.6 reveals a workstation assessment section on the investigation report. It records the human factor program ratings and the behavior-based safety program score for this particular workstation or involved task. We can begin to track, on a log or database, the frequency and severity of critical incidents based on this information. As experience mounts, the organization can have a customized benchmark of scores for the core safety management programs that indicate a correlation of score to critical incidence occurrence. The organization can now place quantitative measures to goal setting for standing problem-solving committees. So, for example, human factor committees can now strive to bring workstation scores to a set level. As experience and problem solving progresses, a mature program now has a very different way to measure success rather than hazard elimination. Hazard elimination, as defined by ANSI B11, as change of function, task or material substitution, is not the only measure for program performance. This measure reflects a well-controlled hazard.

REFLECTION 6.6

1. Why is tracking causation so important to a management system?

AUDITS AS PROACTIVE INVESTIGATIONS

An audit is an investigation for the meeting of a set criteria. Envision a checklist approach to assessment. It can be done in response to an incident when checking compliance-level criteria; however, in the management of occupational safety it is meant to be a proactive investigation. In this safety management system, audits are used primarily as proactive investigations and as a method to identify applicable substandard conditions in an incident investigation.

There are two types of criteria that determine the level of audit. Each core management program and each compliance program will include both levels of audit. The two levels of audit are compliance and effectiveness. Compliance-level audits look for meeting a certain standard or policy. Criteria are developed from a pass-or-fail result directly from the OSHA regulations and from any other policy or practice adopted by the organization. These can include national consensus standards or company-learned practices and policy. Effectiveness criteria identify strong and weak areas for a particular core management program or compliance program. The criteria look to score performance on a scale.

In this safety management system, effectiveness-level criteria follow the Mathis structure of metrics. Level 1 metrics are leading precursors associated with management's duties under the program. Level 2 metrics are leading precursors associated with cultural or operational data, which includes behaviors or observable actions and perceptions. Level 3 metrics are lagging final performance metrics (Mathis 2014).

An audit can also be either internal or external in the terms of personnel who conduct the audit. Internal audits are those performed by a person or team from within the organization. External personnel or teams can also be used to perform audits. Each has its own advantages or disadvantages. Internal audits mainly keep the learned material within the organization. External audits, while sharing learned data to external team member(s), also bring in the experience and knowledge of the outside personnel, which can improve the organization. The safety manager can make or propose when an external audit may be of advantage to the organization.

Compliance audits are measures of substandard conditions. Immediate level of causation consists of two causes: unsafe acts and substandard conditions. Behavior-based safety will address unsafe or undesired acts (covered in chapter 7), so compliance-level audits will address conditions. To alleviate confusion, let me explain that examining behaviors for compliance does occur in compliance-level audits, but in this program, any behavior observation associated with a compliance-level audit will be tracked and addressed in the behavior-based safety program or tracked for a score of compliance in an effectiveness audit.

There is another foundational design to compliance-level audits: ranking levels of criteria to match hazard levels. Level 1 criteria are criteria any authorized safety associate (ideally all employees) can understand and recognize as compliant or noncompliant. Level 2 criteria are criteria competent safety associates with more advanced regulatory knowledge can assess as compliant or noncompliant. Level 3 criteria correspond to criteria that only certain associates with advanced technical training and regulatory knowledge could assess for compliance. This concepts mirrors the categorizing of hazards for hazard recognition.

The concept of categorizing criteria in correlation to competency levels is critical for empowering associates and developing an increased level of commitment. With this concept, it is possible to cull level 1 criteria from a compliance audit and assign workforce associates advanced duties for safety inspections. Furthermore, engineering personnel can examine level 3 criteria. Once safety personnel or competent safety associates assess level 2 criteria, results of compliance can reflect a complete audit.

A perfect example of level 3 criteria in regard to machine guarding is the trouble-shooting of a programmable logic controller (PLC) of a production machine, guarding sensors that indicate enclosed access gates, or light curtain protection device programming. An example of level 2 criteria could be the determination of safety device or setting distance on a power press. Corresponding level 1 criteria could be a loose or missing barrier guard.

Because some criteria or groups of criteria might be used for effectiveness evaluation of a particular program, we must identify criteria used for this purpose and log the results on a database. Table 6.6 is an example for a compliance-level audit criteria heading.

TABLE 6.6   *Compliance Level Audit Criteria*

ITEM #	LEVEL	SOURCE	TRACKING REFERENCE	CRITERIA	PASS (1 PT)	FAIL (0 PTS)

Item numbers make it easy to reference any note or results such as the reason for failure. Tracking reference refers to any level of metric tracking that the criteria may be used to calculate and results in an effectiveness audit for relevant program or policy. Logs allows us to create or use the log as a database for examination of collected data. Criteria are the issues to be assessed and are created directly from regulations, consensus standards, or company policy. They are written in pass-or-fail mode because we are measuring compliance and not effectiveness or overall performance. Failure of any criteria will present a substandard condition that we do not want to determine importance or ranking of in relation to the other criteria.

When compliance audits are completed, the audit can be totaled as a percentage of pass results calculated against possible pass results. These grades can then be tracked by machine or process and topic of audit so that we have comparable scores for compliance by machine, process, or location. A good example is a compliance score for all production machines for machine guarding compliance. Our goal, of course, is one hundred percent on one hundred percent of machines and processes.

Table 6.7 represents a compliance-level audit for general requirements of machine guarding for all machines.

TABLE 6.7 *Machine Guarding Compliance Audit*

ITEM #	LEVEL	SOURCE 1910	TRACKING REFERENCE	REVIEW CRITERIA	CRITERIA	PASS (1 PT)	FAIL (0 PTS)
1	2	212(a)(1)	Level 2 compliance	M 1	One or more methods of protection for the operator. Safety device: _____		
2	2	212(a)(1)	Level 2 compliance	M 1	One or more methods of protection for others in the area. Protection: _____		
3	1	212(a)(2)	Level 2 compliance	M 1	Guards securely fastened to machine and do not create additional hazard.		
4	1	212(a)(3)(ii)	Level 2 compliance	M 1	Point of operation guarding prevents reaching over, under, through during machine movement		
5	2	212(a)(3)(iii)	Level 2 compliance		Special hand tools are required to supplement guarding to handle material in and out of point of operation.		
6	2	212(a)(4)	Level 2 compliance		Barrels, drums, and containers must be fully enclosed with access gate/ door fully interlocked to drive mechanism.		
7	1	212(a)(5)	Level 2 compliance		Exposed blades are fully enclosed unless more than 7 feet above the floor and enclosure must have opening no larger than a 1/2 inch.		
8	1	212(b)	Level 2 compliance	M 1	Fixed location machines shall be securely anchored as to prevent walking or moving.		

1. Why is compliance a criteria of culture?
2. Is compliance a measure of organizational ethics? Why or why not?

BRINGING IN COMPLIANCE RESULTS IN PROGRAM EFFECTIVENESS REVIEWS

All safety programs, core management programs, and compliance-level programs must be reviewed for effectiveness. OSHA even mandates that the compliance program of lock-out tag-out have an annual effectiveness review. Management system standards require management review and continuous improvement efforts. Auditing programs for effectiveness allow managers to focus resources at the right problem areas to improve overall performance.

Chapter 2 covered program effectiveness audits dividing the criteria in management criteria, operational criteria, and status criteria. The goal is to score performance on the criteria and formulate plans to improve. A portion of program effectiveness relies on how compliant the organization is to safety regulations and its own policy. Adherence to compliance-level criteria may be operational for effectiveness reviews, as they may reflect behaviors, and some compliance-level criteria may be considered management criteria for effectiveness reviews since it management's duty to correct substandard conditions. Therefore, each compliance-level criteria used to determine program effectiveness will be labeled as to what effectiveness criteria it corresponds to. At the end of the review period, it would then be possible to have totals for meeting the effectiveness criteria already calculated from performing compliance-level audits.

In Table 6.4, you will notice that each criteria used to calculate a review-level criteria is labeled with the exact number of the criteria where it will be factored into a score for the relevant topic. Since this is a machine guarding compliance-level audit, we would then track the compliance score for the criteria for the entire facility. In other words, if we had one hundred machines to audit for compliance to the general requirements for machine guarding, we could tally a total score for the effectiveness review. For the example, in Table 6.4 we are using five criteria per machine for all one hundred machines to tally a compliance score for the effectiveness review. The management criteria for the effectiveness review for machine guarding would have a possible perfect score of 500, reflected by five complaint points per machine. We can then correlate this raw score to a performance rating of a management-level criteria.

The management-level criteria measure how well management is meeting tis duties under the program. The first management criteria, or M1, for this example would be written as shown in Table 6.8.

TABLE 6.8 *Management-Level Criteria*

ITEM	CULTURE CRITE-RIA	CRITERIA	RESULT
1	Commitment	Machines falling under the general requirements for machine guarding are designed and equipped for compliance to 1910.212	1 2 3 4 5

We could make a determination and establish in the instructions for the audit a resulting review score based on the compliance score. For example, a 5 on the effectiveness review may indicate a score of 495 to 500. The range for a score of 4 may be a little broader. In the beginning we can set these subjectively, and as we obtain more readings over time, our benchmark for excellence may change.

REFLECTION 6.8

1. How does effectiveness auditing contribute to the management process?

POSITIVE AUDITING

We established early on in this chapter that audits are investigations. However, they have formal set of criteria that are assessed. They can be conducted reactively to an incident but are primarily meant for proactive identification of problems. Criteria are typically written and assessed from the aspect of findings causal factors: the negative conditions, acts, factors, or system failures that are associated with facilitating a negative safety occurrence. However, we can also assess for positive precursors at the compliance level as we do in effectiveness audits for management and operational criteria. We can look for issues that are met or compliant with regulations, policy, or best management practices and assess for these rather than look for issues of deficiency.

Using the root-cause model of causation (Bird et al. 2003) as a base template for negative precursors of an incident, we can begin to determine the opposing precursor and assume it is positive. Table 6.7 is an example for producing positive audit criteria. We can begin with some of the standard examples or common examples that may even be present in a particular facility, and, as the program matures, begin to narrow down positive conditions or include the positive conditions that are more critical to success.

TABLE 6.9   *Immediate Level of Causation: Machine Guarding*

SUBSTANDARD CONDITIONS	POSITIVE CONDITIONS	UNSAFE ACTS	DESIRED ACTS
• Open access to electrical • Improper safety device • Lack of protection for others that may be in the area of operations • Inadequate barrier guards • Access gates not interlocked • Enclosures/guards with opening greater than 1/2 inch • Machines not secured to floor	• Electrical boxes with special locking tools for access • Proper safety device for operator • Protection for others that may be in the area of operations • Barrier guards that prevent reaching over, under, or through • Adequate and interlocked access gates • Enclosures/guards with adequate openings (OSHATable O-10) • Machines secured to floor	• Improper use of tool • Working too fast • Working too slow • Failure to follow procedure • Improper wear of PPE • Failure to wear PPE • Improper positioning	• Proper use of tools • Working at an acceptable pace • Paying attention to detail • Following procedure • Proper wearing of PPE • Proper body position

In Table 6.9 we are using substandard conditions from the machine guarding audit as a conceptual example only. We should select the positive criteria from all relevant compliance-level audits and examine for the substandard conditions, unsafe acts, factors, and system failures common to the facility or applied situation in general. Ideally we will only examine for the positive criteria or metrics that are critical to success. Otherwise, we are merely mirroring the results of a compliance-level audit.

Every machine, line, or process will be examined for the positive precursors that are deemed critical from the areas of compliance criteria, design factors, and behaviors for that particular type of job station. We will identify and track in our incident investigations human factors scores, safety behaviors score, and critical conditions to produce a study of sorts that tells us more specifically what the positive precursors are for our facility. For example, if we identify that a certain human factor score from the job station assessment needs to be at a certain benchmark for it to be relatively free from musculoskeletal disorders, then we can work to bring other work stations to this overall benchmark. But perhaps more importantly, if we can look at the assessments that meet that benchmark, there will be certain criteria that stand out as significant to making the job station safe. These then become the specific positive precursors we begin to assess in a positive audit.

> ### REFLECTION 6.9
>
> 1. What role could positive auditing play in managing workplace safety? Isn't auditing for substandard conditions the same concept? Why or why not?

CONCLUSION

Investigational skills are relied on for most of the duties and activities that a safety professional will perform. Successful safety management relies on identifying hazards, risks, and opportunities for making the work place safer; identifying causes; and producing and implementing effective counters. This core function of safety is critical to efficient performance. A safety management system is really a system for organizational learning. Proactive and reactive investigations are at the center of ethical safety leadership. Objective and accurate investigations produce acceptable results for personnel at all levels and justify organizational change.

REFERENCES

Bird Frank E., Jr., George L. Germain, and Douglas M. Clark. 2003. *Practical Loss Control Leadership*, 3rd ed. Duluth, GA: Det Norske Veritas.

Heinrich, H. W. Dan Peterson, and Roos Nestor. 1980. *Industrial Accident Prevention*. New York: McGraw-Hill.

Mathis, Terry. L. 2014. "Common Practice: The Third Level of Leading Indicators." *EHS Today*. ehstoday.com/safety-leadership/common-practice-third-level-leading-indicators.

US Department of Health and Human Services, National Institute for Occupational Safety and Health. 1996. *Current Intelligence Bulletin 57 Violence in the Workplace: Risk Factors and Prevention Strategies* (DHHS Publication No. 96-100). Washington DC: Author.

Figure Credit

CHAPTER SEVEN

Structuring a Behavior-Based Safety Initiative

FOREWORD

Incidents have two immediate causes: substandard conditions and unsafe acts (Bird, Germain, and Douglas 2003). Behaviors are a critical component in preventing incidents. "Behavior based safety is a concept of managing the acts of human participants with the goal of establishing desired behaviors or acts that promote a safer environment, into values, or culture, that produce outcomes" (Dotson 2017, 49). A behavior-based safety program is an effort to understand the difference between designed procedures and actual procedures or steps, it is an endeavor aimed at near-miss recognition, and it is an endeavor to change safety from a priority to a value. It involves associates at all levels of authority and competency. It develops a sense of self-discipline for safety without the direct threat of enforcement. In MacGregor's description, it is a theory Y-type management program because it assumes that resistance to safety is learned, that workers can self-direct, and that the workers are capable of problem solving and adding value to a process. This is important in developing a value that is not yielded in decision making from one situation to another because humans act more moral from ethical beliefs than from the direct following of rules (Zagzebski 2004). Behavior-based safety can develop safety into a virtue that influences safe behavior—a virtue that moves desired behaviors toward acceptance as doing the job right.

Behavior-based safety (BBS) has a conceptual framework that recognizes perceptions, builds from a baseline study, has an authorized level of activity and a competent level of action and administrator action, produces outcomes, and circles back to perceptions.

Objectives

After reading this chapter the learner will be able to do the following:

1. Identify and classify behaviors
2. Perform an initial baseline study of typical behaviors
3. Design behavior-based safety practices on competency level
4. Utilize behavior-based results to help assess risk
5. Design training curriculum based on competency level

LEARNING PLAN

LEVEL	ULTIMATE OUTCOME	CLAIM	LEARNING TYPE	ASSESSMENT
U	Identify and classify behaviors Perform an initial baseline study of typical behaviors	Safety managers must identify behavioral practices common to the facility to design a customized behavior-based safety program.	CT	Learners will observe human behaviors at a crosswalk intersection recording observational data and then categorizing the data based on perceived potential for negative harm.
U	Design behavior based-safety practices on competency level	Behavior-based safety practices are determined by competency level of the observers.	CT	Learners will identify the activity and responsibilities for a behavior-based safety program referencing competency level.
U	Utilize behavior based-results to help assess risk	Risk assessment must include study of correlation between incidents and behavioral performance.	CT	Learners will create a log for tracking incidents in relation to overall behavioral scoring.
U	Design training curriculum based on competency level	Training and participation duties rely on developed competency.	CT	Learners will design training curriculum based on competency levels.

Learning Plan Legend
   Level:
   F: <u>Foundational outcomes:</u> Basic abilities
   M: <u>Mediating outcomes:</u> Progress through a developmental model; interpret, analyze, evaluate progressively challenging claims, arguments
   U: <u>Ultimate outcome:</u> Navigate most advanced arguments and claims

   Type of learning:
   CR: <u>Critical reading:</u> The ability to read, process, and understand the meaning of written information
   IL: <u>Information literacy:</u> Locating and selecting suitable information for a task; evaluating appropriateness/validity of information sources
   AC: <u>Application of concepts:</u> Ability to apply discipline-specific knowledge/skill to tasks/situations important to the discipline
   CT: <u>Critical thinking:</u> Ability to apply a concept to a vague or argumentative claim without a creative leap
   AT: <u>Analytical thinking:</u> Ability to critique/analyze situations using a concept or model
   CA: <u>Creative application:</u> Ability to apply a model/concept in a new way/to an unrelated situation or scenario; involves creative leaps

THE FRAMEWORK

A behavior-based safety program is really a form of action research, or research that has a goal to make change. The study is ongoing and has an ultimate goal of developing safe and efficient behavior associated with work tasks, a virtuosic value.

Many professionals will assert that behavior-based safety is not there to change behavior. This truly means that we are not attempting to force changed behavior from a theory X point of view. Rather than being a carrot-and-stick-type program, proper behavior-based safety programs align themselves with the assumptions of theory Y and apply the assumptions with participation and competency development that allow for team problem solving. Intrinsically stated, the human worker begins to develop value for safety that then becomes a self-motivated habit to get the job done. Foundationally, doing the job right means with safe and efficient actions or the job has not been finished at an acceptable level.

The program begins with a baseline study conducted by the experts. Of course, there are some behaviors we recognize as being common to any task, but to assume that our experience or knowledge of the workstation or of safety behavior in general is sufficient or applicable to each station or task is to allow bias to interfere with objective findings. Behaviors can vary from situation to situation, and slight variances can have very different levels of impact and causation. Alphonse Chapanis, the father of modern ergonomics, established that workstation design can facilitate human error and, conversely, encourage safe and efficient actions. Therefore, each workstation must be approached separately and in enough detail for any findings to be considered valid and lead to effective changes. Changes in work procedures, perception, compliance to procedures, and competency level of involved personnel encourage participation and demonstrate commitment, in other words, culture. Each machine, workstation, assignment area, or other way the site or facility is structured will have postings that identify the desired and undesired behaviors for that station. These are worked into the job hazard analysis, as covered previously in chapter 4. Table 7.1 reflects the format for combining hazard analysis with behaviors.

TABLE 7.1   *Job Hazard Analysis*

| TASK | CRITICAL STEPS | HAZARDS | BEHAVIORS | | COUNTERS |
			DESIRED	UNDESIRED	

Combining these covers the immediate level of incident causation and provides additional hints at basic-level causes of personal factors and human design factors. This methodology gives much more meaning than providing separate postings or only addressing hazards. This is because it is based on Bird's management model of incident causation where the immediate cause level consists of substandard conditions, inadequately controlled hazards, and unsafe acts.

The baseline study begins by conducting initial observations and reviewing observations in comparison with designed and established work procedures. It also includes interviewing the person performing the tasks for purposes of analyzing work procedures for perceived correctness, identification of behaviors common to the task or work station, explanations of any variance between designed procedures and actual procedures, discussion of observed undesired behaviors, and discussion of possible solutions. The baseline study ends when the team of observers and performers can arrive at a base set of work procedures and behaviors relevant to the workstation.

It is then that observations can begin at the various levels of competency and produce more advanced solutions as an ongoing learning process. The box that follows reveals the conceptual framework for the overall behavior-based safety program.

BEHAVIOR-BASED SAFETY
CONCEPTUAL FRAMEWORK

Associates: Frontline, Supervisory, Management
Behaviors
Perceptions
Importance of safety
Company safety efforts
Plausibility of designed work procedures

Baseline Study	
Data collection	Initial observations/procedure review/interview
Reduce data	Identify designed procedures/actual behaviors
Display data	Identify discrepancies
Display data	Identify desired/undesired behaviors
Conclusion	Finalize behaviors

Authorized Level
Peer-to-peer observations: Frontline, supervisory, management
Observer training: Level 1 investigating observer

Competent Level
Observer training: Level 2 investigating observer
Unannounced observation/interview
Competent-to-competent observations: Supervisory, management

Administrator Level
Confirmation observations to authorized/competent level
Tracking measures for success
Adjusting training objectives
Working through system safety analysis for work procedure adjustments

Outcomes
Reinforce positive consequences

Source: Dotson, Blair, Rawlins, and Rockwell 2017.

REFLECTION 7.1

1. Why is BBS described as an action research study?
2. Can all management programs be described as an action research study? Why or why not?

IDENTIFYING BEHAVIORS

Behaviors are observable actions. In this regard, behaviors can be described either quantitatively or numerically, qualitatively, or by words. Both types of descriptors can be used in behavior-based safety, but most will be qualitative in nature. In terms of occurrence, the investigator or observer will not make a determination of motivation for the behavior but may be asked to describe the behavior as creating more potential for negative harm or more potential for producing a less risky environment. We will title these as desired or undesired behaviors.

This does assume that the observer is experienced at judging risk. Specifically, the observer should be trained in analytic induction and in incident causation as well as demonstrated investigational basics such as hazard recognition and causal analysis. Analytic induction is a method of analysis where the incident as a whole is broken down into incremental progression and each step is analyzed for causal factors. It is also important to note that initial observers or experts must also review methodology and models of analysis so findings are consistent.

Initially in the baseline investigation we might have some categories of behavior already identified, and for those outside of this grouping, we can merely record a plain-word description of that observed. Initially we can guess that some behaviors will be common and it is acceptable to list these on an observation instrument or form that collects or records the data. Possible findings are shown in Table 7.2. Here, the observer can add in plain language a descriptor of what was observed, while also being able to record the frequency of the action and adding additional behaviors not anticipated.

RESEARCHER NOTES

Machine/Process/Workstation ID: _____

Date: ___/___/_____   Observer: _____

TABLE 7.2 *Behavioral Notes*

DESIRED BEHAVIORS	UNDESIRED BEHAVIORS
FREQUENCY	FREQUENCY
Proper wearing of PPE	Improper wearing of PPE
Correct PPE	Wrong PPE
Paying attention	Being distracted
Physical balance	Being outside of working zone
Proper working zone	Hurried

The observer may find additional actions from work station to work station, such as use of tools and of other observable facts and described acts that have varying degrees of seemingly interpretable information such as appearing to work frustrated or relaxed. This is why experienced experts with observations or specific knowledge of the workstations must be utilized for the baseline study and group training of methodology must be conducted. Furthermore, even initial observations will include interviews with observed personnel to confirm and validate observations. Once all observers have observed each station, the notes can begin to be analyzed for coding.

BACKGROUND FOR IDENTIFYING BEHAVIORS

There are varying types of observation. Central to a behavior-based safety program is the conundrum of announced versus unannounced observations. In this behavior-based safety program, both will be utilized. To fully understand the implications of these techniques, we will first discuss these as researchers.

Generally, observations are first described as naturalistic or laboratory observations. Naturalistic observations are intended to observe the subject in its natural environment in hopes that they display more realistic behavior. Laboratory observations are those in a controlled setting and typically involve focus on only a few key behaviors. The setting may be controlled to encourage those specific behaviors in specific situations (Jackson 2009). We will consider workstation observations to be naturalistic in regard to work because we are not implementing any type of control or setting influence that is different from what will naturally be encountered during work activity. In a laboratory-type observation in a work setting, we might, for example, in research for the design of a new process, ask a worker to perform a task under a controlled environment, perhaps of dim light or loud noise or under an anticipated environment, to gauge accuracy of work. This type of experiment would then try to project findings to actual work environment in attempt to identify problems and correct them before a prototype was built or before the workstation was put in production.

Next, we must consider understanding nonparticipant and participant investigations. In a participant-type observation, the observer participates in the situation. The observer can get an insider feel to the research in hopes of arriving at more valid findings. Whereas in a nonparticipant observation the researcher does not participate and is out of the immediate environment (Jackson 2009). In workplace observations, participant observations may resemble the safety professional performing all or parts of the task or even being taught the task by the observed person. Nonparticipant observations may resemble watching work activity without performing any part of a task, or even breaking down filmed actions for ergonomic movement.

Although the previously mentioned types of observations can hint at whether the observed party is aware of the observer, they can be disguised or undisguised observations (Jackson 2009). Whether the observed person is aware of the observer can have implications on the findings. If the person observed is aware of the observer, he or she may attempt to behave in a certain manner. But regardless of the description of the observation by type, we must be concerned about the reactivity of the observed and our own expectancy effects or bias.

In this behavior-based safety initiative we will utilize participant and nonparticipant observation techniques along with disguised (unannounced) and undisguised (announced) observations. Many utilize only announced observations for fear of enforcement from a theory X-type application. But our behavior-based safety program is not tied to punishment. In addition, its competency level and non-authority level decides whether the observation is disguised or undisguised. By utilizing these opposite observation techniques, we hope to counter the negative concerns of both and combine results to arrive at a confirmed behavior rating. By incorporating interviews with both disguised and undisguised observations, we will be able to take advantage of insider expertise on the tasks at hand. Furthermore, this helps counteract the observer's own bias or assumptions about the work task or expected behaviors that might be prioritized. By using undisguised observation, we can get a gauge for how well the observed person knows and can demonstrate designed procedures (Note: We did not say proper procedures). By then using disguised observations, we can get a measure for what actually occurs rather than the observed meeting observer expectations. In summary, it is about guaranteeing we have an accurate dataset from which we will analyze and produce counters to problem areas.

CODING THE DATA

For purposes of behavior observation and with most qualitative research, a code is a word descriptor that assigns meaning to the data. Encoding refers to labeling data. Decoding refers to assigning a core definition to the data (Saldana 2009). In an ideal setting we would merely record what behaviors we observe and then try to group or assign them into common meanings. We have already begun the coding process by assigning desired and undesired as the two categories of behavior. Furthermore, we have included some behaviors that we expect or already know will be exhibited. Coding is not exact and is interpretive to the investigator (Saldana, 2009).

Coding can take place from several viewpoints that give meaning to the data. Data can be examined for patterns such as "similarities, differences, frequencies, sequence, correspondence, and causation" (Saldana 2009, 6). Data can be coded several times if necessary for the investigator to narrow the data into categories or realize the best descriptors for the data and the situation. Since the situation here is safe conduct, we can readily identify the categories of desired behavior as being actions that reduce the risk of injury or negative occurrence and undesired behavior or actions that increase the risk of injury or negative occurrence.

However, within these two categories, there can be several subcategories of actions defined by topic of work. This may perhaps be the coding activity that varies with industry, facility, or work situation. In general, behaviors compared against designed procedures are of interest, as well as ergonomic concerns such as physical psychological demands, housekeeping concerns, personal protection equipment usage, and special tool usage. Depending on the situation and the investigator's interpretation, these subcategories may change. For example, if we were working in outdoor conditions and in heat, we may want to examine the area of ergonomic work practice for actions that negate the effects of heat. These might include frequency of hydration and breaks or use of shade and skin protective barriers.

Figure 7.1 is an example of one possible behavior observation card with the situation exampled as general manufacturing.

The baseline study sets an observation instrument for the various situations or workstations present and may prioritize certain actions with the categories of desired or undesired behaviors depending on the application. In theory, the instrument may be different for every workstation. As the program matures, data results may also guide the changes in the observation instrument as problems arise, and counters fix old problems and shed light on other priorities.

Of particular concern with this program and with any managed program is how to turn data, especially qualitative or word descriptors, into numerical data or what is called quantitative data. If you notice on the observational instrument, the observer gives a numerical rating of overall performance. The numerical rating gives meaning to the overall ratio of desired versus undesired actions witnessed per subcategory

Task/Process/Machine		Associate:		Facility:		Date:	
Department:		Supervisor:		Manager:			

Observation:		Desired:	Undesired	Remarks:	Rate 1 to 5
TaskProcedure		☐ Attentive	☐ Too fast		Follows Procedure
1		☐	☐ Distracted		Pace
2			☐ Shortcut		Attention
3			☐ skip step___		
4					
5					
6					
Ergonomic:					Ergonomic Practice:
Physical		☐ Balanced	☐ unbalanced		
		☐ Optimal zone ___	☐ out of zone		
Psychological		☐ relaxed	☐ Frustration		Psychological Stress:
		☐ Motivated	☐ Non Motivated		
Housekeeping		☐ neat area	☐ clutter		Housekeeping:
PPE		☐ Proper wear	☐ improper wear		PPE:
Tool/Equipment Use		☐ Used As designed	☐ wrong use		Tool/Equip Use:
		☐ Wrist/arm neutral	☐ misaligned joint		

Associate Input:	**Self Rating**	**Date Last Training**	**What can EE do better?**	**What can Company do better?**
Task Procedure				
Hazards of Concern				
Ergonomic				
Housekeeping				
PPE				

FIGURE 7.1 BBS Observation Instrument

Source: Dotson 2017.

and category. The observer is asked to keep track of unbalanced movements versus balanced movements and judge overall performance. It is a duty of the administrator to decide, based on specific situational experiences at the facility, any guidance on ratings. In other words, what constitutes a 4 or a 3 rating can be developed and communicated in the training and development of program participants. We can now track these scores and average them to give us overall comparisons from workstation to workstation.

REFLECTION 7.2

1. What would you include in authorized-level and competent-level training plans about conducting behavior observations?

LEVELS OF COMPETENCY

All managed initiatives and programs consist of three levels of competency. These levels determine the duties to be performed based on the level of development or education, training, and experience. Both management and leadership principles demand that personnel only be assigned or evaluated on duties they are prepared to undertake and complete successfully.

The authorized-level associate is the associate who has been educated on the behavior-based program to include what happens, what to expect, how the program is to be used, as well as behavior definition, what to observe, and how observations and subsequent interviews will be conducted. All associates must be developed to the authorized level to participate as an observer.

Authorized-level observations consist of announced peer-to-peer observations. The observer will ask permission or notify the associate that he or she will be observing the associate's work. Since this is an announced observation, the results demonstrate the level to which the worker can demonstrate desired behavior, which includes adherence to established work steps. Once the observer has collected enough data to rate each area of work behavior, procedure, ergonomic, housekeeping, personal protective equipment, and tool usage, the observer will interview the associate.

The interview will review positive findings first, review the overall rating, and then get input from the associate as to what he or she could do to better performance and, most importantly, what the company can do to facilitate better perform. The authorized-level observation is then submitted to the competent level.

The competent-level observer is not necessarily a supervisor or safety professional. Depending on organizational structure and development of competency level with associates, a more mature program would include experienced and successful workforce-level associates as competent-level observers. This is important because most behavior-based safety programs end at the authorized level due to mistrust between workforce and management. The fear is that observations may become attached to disciplinary write-ups. Discipline will not be tied to this behavior-based program either, but because it might discourage reporting and participation in regard to the accuracy of the data.

The competent-level observer developed to the authorized level may have more advanced observational training but truly has experience performing observations and serving on problem-solving committees. This development mirrors the competent-level development from the hazard recognition program in that abatement strategy and problem-solving methodology would be other areas of knowledge and experience developed for inclusion to the competent-level observer rank.

Once an authorized-level observation is completed, a competent-level observer will observe the same workstation or the same workstation and associate to confirm the previous findings. Since this observation is not announced, the measure is more of what occurs than what can be demonstrated. This is important since our goal is to identify issues that influence behavior and find solutions. The observation and subsequent interview is conducted in the same manner. But now we have two measures that probably will differ slightly. We can average the two to get a more complete vision of behaviors at the workstation.

Those at the administrator level have mastered the authorized and competent levels and can now tally results, log the data, and make reports and track the progress of any problem-solving team. These personnel must have the time and management skills to work with numbers. Ideally this is a safety professional at the organization or administrative assistant with the pertinent skills.

Like the other core management programs, the behavior-based safety program must have a data collection log and interface with other programs to factor into measuring the safety aspect of the organization's culture. Metrics of particular interest include a measure of the difference between designed work procedures and actual procedural observation ratings. Figure 7.2 shows a possible example of a data collection log for behavior-based safety. This log collects results and allows for suggestions and committee solutions to be logged. It is in page format due to the large amount of information to be documented. From this log we can add ratings to a research database or continue to average ratings for the areas of observation (procedural, ergonomic, housekeeping, personal protective equipment, and tool usage). Original observation instruments are also maintained per machine or workstation. Each time the sequence of observations progress from authorized to competent, the overall BBS score of the workstation is calculated per area of observation. It is also possible to change the anticipated problems or behaviors if they change. We can now benchmark workstations by BBS score and utilize the rating to prioritize. An important benchmark is to identify the BBS score for the facility and type of workstation that correlates to the occurrence of incidents or injuries.

So far, the program is really about developing safety to the level of virtue in the decision-making process. A virtue is a value that is not yielded by the individual or group during the decision-making process. It does this through participation, empowerment, and converging practical behavior with designed behavior. But BBS is also a program that identifies near-miss trends. A near-miss must be any mistake, purposeful action, or undesired behavior that does not result in an injury, property damage, or other incident that cannot be easily overlooked or not reported. A near miss is still an occurrence, even if we do not know that it occurred. It has a cost as well. The more a particular type of near miss is experienced without direct harm, or in a manner in which it can be overlooked, the more it becomes accepted. The cost may be intrinsic at first, but if all were known, they could be factored into the cost of the damages from repeat near misses that resulted in harm and had to be reported.

Task/Process/Machine			Associate:		Facility:		Date:	
Department:			Supervisor:		Manager:			

Observation Scores					Problem	Committee Solution:	
Procedural							
Authorized Rating:	Follows Procedure				☐ Too fast	Pace:	
	Pace				☐ Distracted	Attention:	
	Attention:				☐ Shortcut	Procedural:	
					☐ skip step___		
Competent Rating:	Follows Procedure						
	Pace						
	Attention:						
Self Ratings Avg.	Follows Procedure						
	Pace						
	Attention:						
Overall BBS rating:	(total of all ratings)						
Ergonomic					☐ Balanced	☐ unbalanced	
Physical					☐ Optimal zone _____	☐ out of zone	
Psychological					☐ relaxed	☐ Frustration	
					☐ Motivated	☐ Non Motivated	
Housekeeping							
PPE					☐ Fail to use	☐ Improper use	
Tool/Equipment Use					☐ Used As designed	☐ wrong use	
					☐ Wrist/arm neutral	☐ misaligned joint	
Record of Incident trends:							
Notes:							

FIGURE 7.2 Workstation Injury Benchmark

Herbert Heinrich first developed the concept of early intervention when he suggested, from his study of workplace injuries during the early twentieth century, a ratio of three hundred near misses to twenty-nine minor injuries for every major injury incurred. The ratio suggests that many near misses will occur before our nemesis, the personal injury. A proper BBS program looks for behavioral trends. But it must also look backward from the reports of incidents that are known. It looks backward to check for effectiveness of the behavior analysis and to identify behavioral trends that had not been identified. We might now create a specific place in the observation instrument for those behaviors relevant to specific workstations or types of workstations.

Each investigation report will have an area for workstation assessment that can record the overall BBS score. For management purposes, each workstation should have its own log or file of findings. This log serves to document BBS program findings, human factor program findings, and results from any risk analysis or investigational findings relevant to the workstation. Figure 7.3 serves as a possible example of this database.

Year 20xx	Workstation	Type	BBS Score	Usability Satis/Efficiency	Injuries	Injury Cost	Compliance Scores			
							Guarding	LOTO	Electrical	WorkSurface
	AS 1	Robot welder	xx	xx	x	xxx.00	8			

FIGURE 7.3 Risk Analysis Database

REFLECTION 7.3

1. What are the authorized-level observations measuring? What are the competent-level observations measuring? By comparing the two, what are we identifying?
2. What are we assuming about workforce-level associates if we do not develop them into competent-level observers?
3. How does the program define near misses?

BBS EFFECTIVENESS REVIEW

Review for effectiveness is a key concept for the proper management of any safety initiative. Tables 7.3–7.5 suggest some key criteria for reviewing effectiveness of a behavior-based safety initiative.

TABLE 7.3 *Management-Level Criteria*

ITEM	CULTURE CRITERIA	CRITERIA	RESULT
1	Participation	All workstations have competent-level observations completed at an adequate rate to keep pace with authorized-level observations and provide a BBS composite score.	1 2 3 4 5
2	Commitment	The organization has approved resource requests from committee recommendations adequate to support the program.	1 2 3 4 5
3	Competency	The organization provides training and ensures experience to associates at a level that allows adequate participation.	1 2 3 4 5
5	Compliance	The company adheres to and enforces BBS procedures.	1 2 3 4 5

TABLE 7.4 *Cultural/Operational-Level Criteria*

ITEM	CULTURE CRITERIA	CRITERIA	RESULT
1	Perception	Workforce associates perceive the program to be important.	1 2 3 4 5
2	Participation	Authorized associates initiate authorized observations adequately.	1 2 3 4 5
3	Participation	Competent-level associates initiate competent level observations adequately.	1 2 3 4 5
4	Perception	Workforce associates perceive management from viewing the program as important.	1 2 3 4 5
5	Perception	Workforce-level associates perceive the company as following policy.	1 2 3 4 5
6	Perception	Management-level associates perceive the program to be important.	1 2 3 4 5
7	Perception	Management-level associates perceive the company as following policy.	1 2 3 4 5

TABLE 7.5 *Status/Performance-Level Criteria*

ITEM	CULTURE CRITERIA	CRITERIA	RESULT
1	Performance	Percentage of workstations with composite BBS scores within sixty days: _____	1 2 3 4 5
2	Performance	Percentage of implemented countermeasures resulting from the BBS program: _____ Number of requested countermeasure projects/actions: _____	1 2 3 4 5
3	Performance	Number of injuries resulting from/involving unsafe acts: _____ Percentage of total organizational injuries resulting from/involving unsafe acts: _____	1 2 3 4 5
4	Performance	Average workstation BBS composite score: _____	1 2 3 4 5

REFLECTION 7.4

1. What goal does each performance metric seem to measure and what is its significance?

CONCLUSION

The goal of safety leadership is to develop safety as a virtue or unyielding value in the decision-making process of the organization as a group and for its individual members. The values instilled as unyielding, and to the degree that they are not yielded, define the organization's culture.

Behavior-based safety (BBS) is a critical component to developing a culture that reflects workplace safety as a value. Conditions and behaviors combine to create the immediate level of incident causation (Bird et al. 2003). Safety is most successful conducted from the aspect of transformational leadership (Ruggieri and Abbate 2013). Transformational approaches begin from the assumption that the leader exists to serve the needs of the subordinates in a way that furthers the needs of the whole while developing the growth of the organization's members. This leadership style must prevail in any behavior-based safety initiative because of the assumptions that the policy or program makes about the workers. If behavior-based safety programs were mere disciplinary inspections or enforcement efforts, you may picture typical highway safety enforcement efforts, they would be conducted as theory X-type programs. These types of programs are the carrot-and-stick incentives that assume workers are lazy and will not make self-directed efforts to perform correctly and therefore must be coerced into behaving correctly (MacGregor 2006). But a behavior-based safety initiative is a theory Y-type initiative. It assumes that workers will make an effort to work safe, that, if committed, workers will self-direct proper behaviors so self-actualization or professional development can be satisfied, at least partly, through meeting safety objectives and that workers are capable of solving safety problems.

The examination of BBS from the view of assumptions reflects reasons why typical BBS programs in companies are lackluster in performance. If management is predominately theory X in its approach, the carrot-and-stick coercion has created an atmosphere of distrust that is difficult to overcome when managing a business concern such as safety, which must be approached from a theory Y approach to policy and lead from a transformational style.

The BBS program begins with an observational study that reflects the differences between workstations and duty assignments, then develops into an observational study conducted by workforce-level associates who have been professionally developed to participate at the entry level and at competent levels. Commitment is developed from participation by conducting two levels of observational study in the form of action research aiming to make a change. The change is not aimed primarily at the behavior of the observed worker, but at identifying near-miss activity and solving the problem of a difference in designed actions and actual behaviors. Much of the discrepancy between the two must point to an examination of workstation design, covered in the human factors initiative.

REFLECTION 7.5

1. MacGregor (2006) teaches us that when a need is not fulfilled it then becomes a motivator of action and we can predict even subversive action to policy that does not fill human need. Reflect to chapter 1 and consider Maslow's hierarchy of needs in relation to a BBS program led from a view of theory X and then from an approach using theory Y. What predictions of behavior can you foresee from the two opposing approaches?

REFERENCES

Bird Frank E. Jr., George L. Germain, and Douglas M. Clark. 2003. *Practical Loss Control Leadership,* 3rd ed. Duluth, GA: Det Norske Veritas.

Dotson, Ron. 2017. "Behaviors Equal Culture." In *Principles of Occupational Safety Management,* ed. Ron Dotson, Troy Rawlins, Earl Blair, and Scott Rockwell. San Diego, CA: Cognella.

Jackson, Sherri, L. 2009. *Research Methods and Statistics; A Critical Thinking Approach,* 3rd ed. Belmont, CA: Wadsworth.

McGregor, Douglas. 2006. *The Human Side of Enterprise Annotated Edition.* Annotated by Joel Cutcher-Gershenfeld. New York: McGraw-Hill.

Ruggieri, Steffano, and Costanza Scaffidi Abbate. 2013. "Leadership Style, Self-Sacrifice, and Team Identification." *Social Behavior & Personality: An International Journal* 41, no. 7: 1171–1178. doi:10.2224/sbp.2013.41.7.1171

Saldana, Johnny. 2009. *The Coding Manual for Qualitative Researchers.* Thousand Oaks, CA: SAGE.

Zagzebski, Linda. T. 2004. *Divine Motivation Theory.* New York: Cambridge University Press.

Figure Credit

CHAPTER EIGHT

Structuring a Human Factors Initiative

FOREWORD

A human factors program is where the worker and machine meet. The initiative described in the chapter is derived from the philosophies of Alphonse Chapanis, the father of modern ergonomics. Chapanis approached teaching human-centered design from the view that a human factors professional must take general design principles and user needs and incorporate them into specific work system design specifications for the situation. His approach differed in that rather than attempt to produce general design laws that would be applied universally by designers, he approached how the principles of human design could be approached in the various stages of development for individualized application. His methodology was to study how engineers go about their business and figure out what human factor input they need at each stage of development (Chapanis 1996). This methodology fits the role of a safety professional working with a cross-functional design team in the development of new workstations or in the modification of existing workstations.

A human factors program is truly a partnership with an organization's cross-functional departments. Its success is a reflection of the commitment of these departments working as a team of professionals fulfilling each department's role. The better the teamwork, the better the results are reflected in reducing resource usage for training, maintenance, user support, errors, and negative occurrences in general. Efficiency is directly proportional to the commitment for a human factors program. Many of an organization's basic needs are met within its human factors program. Human resources, for example, must base recruitment, selection, and placement on the needs of the work environment. Training directors must base training curriculum on the needs of the work environment as well. Maintenance managers can use preventive maintenance scheduling to help manage maintenance resources. Each cross-functional department has specific duties and then each relies on the other to identify information for the management of the departmental function.

This chapter will lay the foundational knowledge and then identify the specific duties that safety must perform and the contributions safety must make during the stages of development or modification and during routine workstation assessment. Chapanis gave the safety manager his or her role in a human factors program: Provide the needed human factor inputs to the designers (1996). This program accomplishes this task. Specifically, it will address a system safety analysis policy, safety problem-solving methodology, and workstation assessment.

Objectives
After reading this chapter the learner will be able to do the following:

1. Define human factors
2. Identify the foundational principles of human-centered design
3. Apply the concept of usability to workstation ergonomic assessment

LEARNING PLAN

LEVEL	ULTIMATE OUTCOME	CLAIM	LEARNING TYPE	ASSESSMENT
F	Define human factors	Safety professionals must be capable of defining and describing the concept of human factors.	CR	Learners will be able to define the concept of human factors by applying the PEAR concept of human factors.
F	Identify the foundational elements of human-centered design	Safety professionals must be able to identify key principles of design to perform their duties when assigned to system safety analysis committees.	CR/CT	Learners will be able to recognize critical elements of human-centered design.
U	Apply the concept of usability to workstation ergonomic assessment	Applying the concept of usability to ergonomic assessment provides a comprehensive approach to human factors.	AC/CT	Learners will perform a usability assessment on a common workstation available to them. (vehicle)

Learning Plan Legend
Level:
F: <u>Foundational outcomes</u>: Basic abilities
M: <u>Mediating outcomes</u>: Progress through a developmental model; interpret, analyze, evaluate progressively challenging claims, arguments
U: <u>Ultimate outcome</u>: Navigate most advanced arguments and claims

Type of learning:
CR: <u>Critical reading</u>: The ability to read, process, and understand the meaning of written information
IL: <u>Information literacy</u>: Locating and selecting suitable information for a task; evaluating appropriateness/validity of information sources
AC: <u>Application of concepts</u>: Ability to apply discipline-specific knowledge/skill to tasks/situations important to the discipline
CT: <u>Critical thinking</u>: Ability to apply a concept to a vague or argumentative claim without a creative leap
AT: <u>Analytical thinking</u>: Ability to critique/analyze situations using a concept or model
CA: <u>Creative application</u>: Ability to apply a model/concept in a new way/to an unrelated situation or scenario; involves creative leaps

THE FOUNDATIONS OF HUMAN FACTORS

Principle Theory

A human factors program is where human well-being and performance (concurrent and equal concepts) meet the technology deployed in the work station. The term "human factors" is used to describe the management initiative because, in practice, the term "ergonomics" is used in slang to refer to the direct management of human musculoskeletal injuries and conditions as a main focus of physical demand. Ergonomics and human factors are truly the same concept. Ergonomics refers to the study of the work station. Central to human factors management or ergonomics is the concept of human-centered design. Human-centered design is the concept of applying a body of knowledge about human anthropometrics, or ability, limitations, and characteristics, to the design of work systems where humans and machines interact to maximize production (Chapanis 1996). We define production as quality plus safety; otherwise, it's inefficient. Simply said, it's building work stations around the needs of human counterparts.

Human factors become central to safety because they address the basic level of causation that precipitates immediate causes and are a window to root causes. In the Bird model of causation, the immediate level, consisting of unsafe acts and substandard conditions, is the most direct and immediate causal factor that allows for a negative incident or occurrence with increased risk for negative consequences to occur. The basic level of causation consisting of job factors and personal factors allows the immediate level to develop or increase the potential when an immediate level factor is introduced. The root level of causation consists of management-level system failures or overlooked issues that were not present or could have prevented the basic level from developing (Bird, Germain, and Douglas 2003).

Examining the basic level of causation's personal factors typically refer to conditions that influence behavior through stress. Common examples include fatigue, being in a hurry, feeling overwhelmed cognitive difficulty, frustration, willingness to please or fit in, lack of experience, lack of skill, or life issues that distract from work due to stress, such as financial problems, addictions, or returning from vacation. Job factors can refer to conditions that must be overcome but are not substandard, such as the difference of operating a tractor on flat ground versus on a slope, or operations in dry climates versus wet, but also include job complexity, tool and equipment availability, and overall physical demand. Human factors are central to these but may not be readily apparent without direct knowledge of human-centered design principles. This is why I cull human-centered design principles at the basic level when teaching the application of incident causation. A good example is the stress that noise can produce in a work environment, even when it meets OSHA standards and is not considered substandard. Depending on the complexity of the job task, noise can multiply stress to the point of facilitating human error. Another good example is how well the job tasks require attention to detail without overloading the worker to the point of facilitating a mistake.

Root causes are the management system failures or omissions that allow basic-level causes to exist. A clear example is the lack of adherence to cross-functional team review for workstation changes. Numerous examples exist common to general manufacturing. For example, in one instance a maintenance technician or his manager may make a decision to increase weld amperage to counter a quality issue of an incomplete weld. But without cross-functional consideration of the consequences of increased amperage, the correction may create other problems that are just as critical. It might warp the material due to increased temperature of the steel and result in an increased hazard to the worker that may not be within the worker's control.

A foundational theory of modern safety management originated with Herbert Heinrich, the father of modern safety management, in the 1930s when he published his ratio of near misses to minor injuries to major injuries. The ratio 300:29:1 laid the foundation for a concept called "early intervention": If a safety manager could identify trends of near misses early enough and correct them, a major injury could be adverted. This concept is not unique to safety at all. It is business. The sooner a problem is recognized and corrected, the more efficient the production process. This means profit. A human factors program operates on this principle first and foremost. Therefore, a human factors program will be an ongoing action research study aimed at identifying the human-centered design principles relevant to the organization's situation and at providing this information to workstation design teams for a continuous improvement effort of human performance. We will do this by conducting full usability assessments of work stations and individual tasks. Before arriving at application of usability, a concept developed by Chapanis (1996) as a meeting of effectiveness, efficiency, and satisfaction, we must first develop a more specific definition of human factors.

The PEAR Conceptual Map

The PEAR model developed by Dr. W. B. Johnson and Dr. M. E. Maddox (2007), is an acronym that stands for people, environment, actions, and resources. It provides a more comprehensive and easily understood visual to what encompasses the broad field of human factors. Figure 8.1 represents a possible conceptual map of PEAR.

This conceptual map demonstrates the model's complexity as it clearly depicts environments having an all-encompassing impact on all the subcategories of human factors while centering on the human being. Johnson and Maddox subdivide people with four categories for assessment: physical factors, physiological factors, psychological factors, and psychosocial factors. Table 8.1 reflects the assessment categories and gives example issues and meanings.

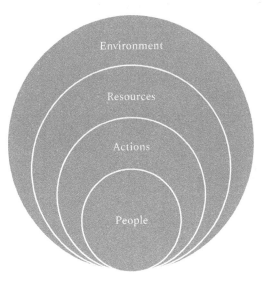

FIGURE 8.1   PEAR Model

Source: Dotson 2017

TABLE 8.1   *Johnson and Maddox (2007) Assessment Categorization*

COMPONENT	CATEGORY	TYPE/DEFINITION	EXAMPLE
People	Physical	Anthropometric	Size and fit of worker Flexibility required Strength required
People	Physiological	Subconscious body control	Hearing Sight Smell Respiration
People	Psychological	Mental	Cognitive capacity Knowledge Experience Perceptions Decision making
People	Psychosocial	Social interaction	Group behavior Relationships

The environmental component of PEAR contains two subcategories for assessment: physical environmental conditions and organizational conditions. Table 8.2 provides a visual explanation for this component.

TABLE 8.2   *Environmental Component of PEAR*

COMPONENT	CATEGORY	TYPE/DEFINITION	EXAMPLE
Environment	Physical conditions	Received physically	Temperature Lighting Spacing Surface hardening
Environment	Organizational conditions	Received mentally/stress	Overall culture Labor/management relations Relationship dynamics

Source: Johnson and Maddox 2007.

The actions component is truly the observable behaviors or interactions at the workstation and includes the interactions between the human and machine counterpart (Johnson and Maddox 2007). Much of this is covered in the behavior-based safety program, addressed in chapter 7, that focuses on work procedures. For this part of assessment in the human factors program, we will assess assignment of general tasks, or the role of humans and machines, to match relevant strengths to the proper counterpart utilizing the Meister human-machine interaction model. Table 8.3 reflects the actions component as developed by Johnson and Maddox.

TABLE 8.3 *Actions Component of PEAR*

COMPONENT	CATEGORY	TYPE/DEFINITION	EXAMPLE
Actions	Machine and human Interactions	Observable actions	Input Observation Pace Task steps Control feedback

Source: Johnson and Maddox 2007.

The resources component of PEAR consists of soft and hard assets. Note that there are numerous possible specific examples. Table 8.4 reflects the resources component.

TABLE 8.4 *Resources Component of PEAR*

COMPONENT	CATEGORY	TYPE/DEFINITION	EXAMPLE
Resources	Hard assets	Tools Fixtures Guides Controls	Drift pins Pliers Light curtains PPE
Resources	Soft assets	Training Education	On-work process Feedback meaning

Source: Johnson & Maddox 2007.

HUMAN-CENTERED DESIGN

Human-centered design means that the components from the PEAR model focus on fitting the human from the four frameworks of human-centered design: safety, performance, comfort, and esthetics (Boy 2011). These frameworks will become important as usability assessments are implemented as a management tool for human factors.

Chapanis established the five elements of human-centered design: personnel selection, personnel training and development, machine design, job design, and environmental design. He also established objectives that give insight to the value of proper human factors management and can be used to structure metrics that show program success. The following list reflects some of the more important objectives for human factors:

1. **Basic operational objectives:**
    Reduce errors
    Increase safety
    Improve system performance

2. **Objectives for RMA: Reliability, maintainability, availability:**
    Increase reliability
    Improve maintainability
    Reduce personnel requirements
    Reduce training requirements

3. **Objectives for users:**
    Improve working environment
    Reduce fatigue/physical/mental stress
    Increase comfort
    Reduce boredom
    Increase ease of use
    Increase user acceptance/satisfaction

4. **Other objectives:**
    Reduce loss of time/equipment
    Increase economy of production

This list of more important objectives provides a clear indication that a human factors program is also a management program for other cross-functional departments such as maintenance, engineering, and human resources (Chapanis 1996).

Chapanis is perhaps more recognized as developing usability as concept for assessing human-centered design. In the concept, usability increases as effectiveness, efficiency, and satisfaction merge. For purposes of teaching occupational safety management or for safety management practices, I prefer to cull safety efficiency from effectiveness and efficiency to make usability more clear for practicing safety professionals. Figure 8.2 provides a visual model for this concept.

Effectiveness really means the completeness to which a user achieves a goal. In a work system, it is the completeness of completing procedural steps or critical tasks that in turn combine to complete a job. Efficiency is the accuracy for completing those critical tasks or goals. Satisfaction is the degree to which a user accepts the interaction and features. Satisfaction includes personal preferences and likes and dislikes as well.

Analyzing effectiveness in a usability assessment begins with examining the assignment of roles and duties between the machine and the human components of the work system. The Meister model of human and machine Interaction has three components:

FIGURE 8.2   Usability
Source: Dotson, 2017.

1. The human system
2. The machine system
3. The environment that surrounds both (Meister 1971)

The human side of the system is organized into the sensory function, or how information is detected, the cognitive function, or human decision making, and the human musculoskeletal function, or the actual movement and input from the worker (Meister 1971).

The machine side of the system is a mirror of the human side. The machine input component is how the machine receives information. The machine cognitive component is the programmable logic controller, or processing unit that controls machine programming, and the machine display component is how the machine relays information or feedback to the users of the system (Meister 1971).

The environment that surrounds both has enormous impact on the interaction or usability of the system. Referring to the PEAR model of human factors, environmental conditions have both physical and mental impacts on the human component (Johnson and Maddox 2007).

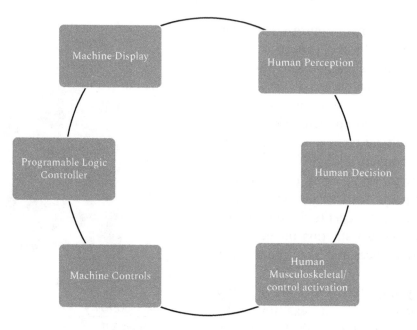

FIGURE 8.3 Meister Model of Human and Machine Interaction

Source: Meister 1971.

Figure 8.3 demonstrates the Meister model of human and machine interaction. System safety analysis is really a control for change. In other words, before a new workstation is placed in service or an existing workstation is modified, a cross-functional team must examine the system for early identification of problems and for competent management of the work process. Organizations can have different titles or steps to new launch or modification, but generally they could be listed as the following:

1. **Concept stage**: First presentation of need/idea/initial concepts are considered.
2. **Prototype stage**: Initial design(s) are brought into being and are ready tested.
3. **Initial testing stage**: The machine or process is initially trialed using engineering personnel.
4. **Testing stage**: Test runs for production are initiated with select line associates.
5. **Initial production**: Production is completed with initial line associates on limited days and/or times.
6. **Production**: The new process or machine is in a normal rotation of full scheduling, and production is possible with all line associates.

Each step has several duties that the safety professional must perform before the project progresses. As soon as possible, the safety professional must identify the interaction between the machine and worker and assess it for effectiveness.

To assess effectiveness, we will begin to examine whether the roles assigned to the human and to the machine match respective strengths. Table 8.5 depicts humans and machines' strengths.

TABLE 8.5 *Human and Machine Strengths*

HUMAN STRENGTHS	MACHINE STRENGTHS
• Sensory functions	• Alertness
• Perceptual abilities (stimulus, generalization, abstract concepts)	• Speed/power
	• Sensor detection outside of human range
• Flexibility (able to improvise)	• Routine work
• Judgment	• Computation
• Selective recall	• Short-term memory
• Learning	• Simultaneous activities

Source: Fitts and Posner 1979.

We initially begin assessment by an inventory of how the machine and the human will function in each step of interaction: how the machine will display information to the user and then how the user is to perceive this information. We will rate how well this fits the strengths of the human user based on the strengths listed and on if the human user can complete the required action. Table 8.6 demonstrates a visual model of the type of assessment that will begin our effectiveness rating.

TABLE 8.6 *Assessment of Human/Machine Interaction*

Human sensory	What is sensed? How is it sensed?	Human limits: Human needs:	Machine Display	Types? Specifics?
Human cognitive	What must be known? What level/requirements of mental ability are needed? What is processed?	Human limits: Human needs:	Machine CPU	Type of data processed: Safety specific programming:
Human musculo-skeletal	What movements? Body parts are involved? Load?	Human limits: Human needs:	Machine input:	Control types: Feedback:

Source: Dotson 2017.

In the example assessment, we note the type of display or action and how it occurs. The notes section is for specific notes of how this occurs. For example, with machine display, if the monitor displays feedback in a certain color or certain display of control, such as a tachometer, we can make specific notes that indicate details that may need assessed. We may then be able to identify any weaknesses or strengths.

After we have inventoried the functions and roles based on human and machine interaction, we can inventory and list the exact functions or task steps that will be measured for effectiveness. This can begin as we initially design the work procedures or as we modify them for existing workstations. In other words, we inventory the necessary completed actions that humans will have to perform, along with the steps that the machine or workstation will complete. For example, let's look at a simple workstation where a worker will take a stamped piece of steel, place it in a spot welding machine, activate the machine controls, retrieve the finished part, and place the part in a parts bin. If we inventory these critical steps and then combine these steps with the same critical actions that the machine or workstation will perform, we can assess whether the action is best for the human or the machine and we can then measure if the completed action steps are finished.

Table 8.7 depicts a possible assessment for critical workstation steps.

TABLE 8.7 *Workstation Step Assessment*

HUMAN TASKS	EFFECTIVENESS	STRENGTH GAPS	MACHINE TASKS	EFFECTIVENESS	STRENGTH GAPS
Action 1 Obtain blank steel	Completed: Y N Difficulties:	Minimal: 1 Moderate: 2 Severe: 3 Catastrophic: 4	Action 1 Recognizes blank availability Signals PLC for blank availability	Completed: Y N Faults:	Minimal: 1 Moderate: 2 Severe: 3 Catastrophic: 4
Action 2 Place blank in welder	Completed: Y N Difficulties:	Minimal: 1 Moderate: 2 Severe: 3 Catastrophic: 4	Action 2 Recognizes part placement Signals PLC proper placement	Completed: Y N Faults:	Minimal: 1 Moderate: 2 Severe: 3 Catastrophic: 4
Action 3	Completed: Y N Difficulties:	Minimal: 1 Moderate: 2 Severe: 3 Catastrophic: 4		Completed: Y N Faults:	Minimal: 1 Moderate: 2 Severe: 3 Catastrophic: 4

When human tasks are assigned within human strengths and can be completed, efficiency assessment can begin. Efficiency is truly an examination for error and error rates. But it is also an examination of time needed to train and reach mastery of the individual steps and, in turn, overall mastery of the workstation. Therefore, we must define mastery in terms of what is an acceptable amount of error in a timeframe. Additionally, we should consider the severity of the errors that do occur, such as in Table 8.8.

TABLE 8.8 *Efficiency and Error Assessment*

HUMAN TASKS	EFFICIENCY	ERROR SEVERITY	MACHINE TASKS	EFFICIENCY	STRENGTH GAPS
Action 1 Obtain blank steel	Errors: Attempts: Timeframe:	Minimal: 1 Moderate: 2 Severe: 3 Catastrophic: 4	Action 1 Recognizes blank availability Signals PLC for blank availability	Errors: Attempts: Timeframe:	Minimal: 1 Moderate: 2 Severe: 3 Catastrophic: 4
Action 2 Place blank in welder	Errors: Attempts: Timeframe:	Minimal: 1 Moderate: 2 Severe: 3 Catastrophic: 4	Action 2 Recognizes part placement Signals PLC proper placement	Errors: Attempts: Timeframe:	Minimal: 1 Moderate: 2 Severe: 3 Catastrophic: 4
Action 3	Errors: Attempts: Timeframe:	Minimal: 1 Moderate: 2 Severe: 3 Catastrophic: 4		Errors: Attempts: Timeframe:	Minimal: 1 Moderate: 2 Severe: 3 Catastrophic: 4

Assessing efficiency means that human workers are interfacing with the prototype system. Initially this is completed by the designers or engineering personnel. But in the testing stages of development, the user will switch to select workforce associates. This means that personnel selection must be addressed.

Since this is a workstation in a particular facility and production line or process, the sample of workforce-level associates must come from the group that will be utilizing the system. Some have a tendency to utilize their most reliable workers. This is understandable from the aspect of having them available on a consistent basis at specified times and days for testing. But to get more credible results, or results that will likely be arrived at during normal production processes, we should consider a certain set of demographics or descriptors of the initial users that will more closely represent the diversity of the demographics of users once the system is routinely being used and scheduled in regular production. You should consider demographics such as the following:

- Education level or cognitive ability
- Experience at the facility
- Training
- Familiarity with similar systems/technology

Ideally, initial testers that will be used to identify and correct initial problems should have the high, middle, and low range of ability. Since testing in a work setting usually does not involve a large sample of workers, utilizing random samples may not give credible results as compared to using high, middle, and low ranges of these demographics.

A minimal error might be defined as any discrepancy from intended action that interferes or slows down completion of the action. A moderate error disrupts production of the system. Severe errors require repair. Catastrophic errors produce reasonable probability of human injury. As we identify errors on the human or the machine side of the interaction model, we can begin to critically assign the steps and identify hazards and near misses. The human-machine interaction model can also aid engineering in determining the criticality of some components and in producing fault trees for the system.

Early identification as a principal of safety management is visible in error assessment as proactively as possible. During the process of creating new or modifying existing work stations and tasks, possible failures must be assessed and overcome or planned for. Each cross-functional component of human factors teams must utilize its respective specialties for identifying possible failures and eliminating as much as it can during the developmental stages. Table 8.9 reveals one such form for documenting and assessing failures.

TABLE 8.9 *Failure Assessment Form*

FAILURE	PROBABILITY	SEVERITY (OPERATIONAL)	EXPOSURE (HUMAN)	CAUSES	CONSEQUENCES
Failure: Critical failure Y □	Rare □ Uncommon □ Frequent □	Little consequence □ Minor □ Moderate □ Severe □	**Number:** Low: 0–1 □ Medium: 2 □ High: 3 or more □ Catastrophic: □ **Severity:** Low: Medium: High: **Frequency:** Low: Medium: High:		

Once possible failures have been identified and assessed, the failure can be mapped. The butterfly method of mapping utilized in the investigational process maps incremental events or observable occurrences leading up to a point or no return or critical event and then the resulting and possible events. Butterfly-mapped failures are troubleshooting charts. They reveal causes on the left side of each failure and show consequences on the right side of the map. The concepts are very similar in nature. This method of mapping allows for causal analysis at each cause and consequence. Failures include human acts as well as mechanical and electrical failures. Each cross-functional division can proactively work to eliminate failures at the workstation in a systematic and proactive manner. Figure 8.4 shows a concept of butterfly mapping applied to all failures.

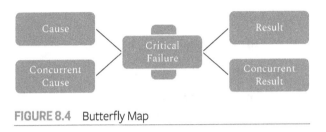

FIGURE 8.4 Butterfly Map

Failures that are not eliminated but controlled now have a systematic method for quantifying risk. If numerical values are assigned to the risk assessment matrix presented in Table 8.9, the organization can adhere to the American National Standards Institute (ANSI) concept of acceptable risk. This means if a suggested control reduces the risk of the failure, the control should be developed and placed (ANSI 2008). This method also provides a measure of the human factors program success. Failures eliminated prior to any release of the new workstation or any modification to routine production would indicate positive performance.

Satisfaction is the third formal component to usability and includes user preferences, perceptions, likes, and dislikes. This is conducted through consultation and then participative surveying of the users. During the initial testing of the system observers can interview for specific likes and dislikes of the user. As this process progresses and workers become more familiar with the system, they can complete surveys independently of an observer. The box that follows exemplifies one possible satisfaction survey. These surveys can and should be tailored to ask about different sections or components of the system. In general, we could categorize the areas of overall design, controls and interface, human fit and comfort, maintainability, and likability. These categories will match the categories that the safety professional will assess and have a committee assess.

User Satisfaction Survey Template

User: Age:

Years with company: Years using similar systems:

Technical competency: New   Apprentice   Journeyman/Experienced   Instructor

Safety competency:

Workstation: _____

Total usability: / 360

Overall Design Satisfaction: Total: /50

Directions: Circle the appropriate rating.

The workstation is simple to understand.

1 2 3 4 Simple   Not applicable

Preoperation checks are simple to remember.

1 2 3 4 Simple   Not applicable

Preoperation checks are easy to perform.

1 2 3 4 Simple   Not applicable

Basic operations are simple to perform.

1 2 3 4 Simple   Not applicable

Shutdown procedures are simple to perform.

1 2 3 4 Simple   Not applicable

Controls are similar to other machines I operate.

1 2 3 4 Similar   Not applicable

The workstation is interesting to me.

1 2 3 4 Interesting   Not applicable

The work station is organized.

1 2 3 4 Organized   Not applicable

The pace of work is too fast.

1 2 3 4 Simple   Not applicable

The pace of work is boring.

1 2 3 4 Boring   Not applicable

Control Interface: Total: / 35

Directions: Circle the appropriate rating.

The controls fit me.

Fit   4   3   2   Do not fit   Not applicable

The controls have the effect I expected.

Expected output   4   3   2   Output not expected   Not applicable

The controls are familiar to me.

Familiar   4   3   2   Unfamiliar   Not applicable

The controls are easily learned.

Easily learned   4   3   2   Difficult to get used to   Not applicable

Controls are readily visible.

Readily visible   4   3   2   Controls are not visible   Not applicable

I can identify the controls in at least two different ways.

Redundant controls   4   3   2   Not applicable

Unintended activation of the machine(s) in the workstation is difficult.

Difficult   4   3   2   Easily activated accidentally   Not applicable

Human Comfort: Total: / 60

Directions: Circle the appropriate rating.

The workstation is too bright.

Too bright   2 3 4 Well-lit   Not applicable

The workstation is too dark.

Too dark   2 3 4   Well-lit   Not applicable

I can see and read controls and screens.

Difficult   2 3 4 Easy to read   Not applicable

The workstation is free of blind spots due to shadows.

Shadows prohibit vision   2 3 4 No shadows   Not applicable

Lighting in the workstation flickers.

None   2 3 4 Frequently   Not applicable

My actions stay between mid-chest line and waistline.

Out of optimal zone 2 3 4 In optimal zone   Not applicable

I often twist in the workstation.

Often 2 3 4 Rarely   Not applicable

I often bend over in the workstation.

Often 2 3 4 Rarely   Not applicable

I often bend my neck down at the workstation.

Often 2 3 4 Rarely   Not applicable

I often bend my neck looking upward at the workstation.

Often 2 3 4 Rarely   Not applicable

I often twist my neck for extended time at the workstation.

Often 2 3 4 Rarely   Not applicable

I am tall enough to reach all controls of the workstation.

Tall enough 2 3 4 Too tall

Physical Demand: Total: / 75

My forearms feel bad after working at the workstation.

No 4 3 2 Yes

My arms feel bad after working at the workstation.

No 4 3 2 Yes

My wrists feel bad after working at the workstation.

No 4 3 2 Yes

My shoulders feel bad after working at the workstation.

No 4 3 2 Yes

My feet feel bad after working at the workstation.

No 4 3 2 Yes

My legs feel bad after working at the workstation.

No 4 3 2 Yes

My back hurts after working at the workstation.

No 4 3 2 Yes

My neck hurts after working at the workstation.

No 4 3 2 Yes

My fingers hurt after working at the workstation.

No 4 3 2 Yes

I feel vibration at the workstation.

Does not affect me 4 3 2 Creates discomfort

It is too hot at my workstation.

Does not affect me 4 3 2 Creates discomfort

It is too cold at my workstation.

Does not affect me 4 3 2 Creates discomfort

The temperature changes too often at my workstation.

Does not affect me 4 3 2 Creates discomfort

It is too loud at my workstation.

Does not affect me 4 3 2 Creates discomfort

There is too much motion at my workstation.

Does not affect me 4 3 2 Creates discomfort

Physical Symptoms: Total: / 30

I wake up at night with wrist or forearm pain after working at the workstation.

Never 4 3 2 Often

My fingers go numb after working at the workstation.

Never 4 3 2 Often

My hands go numb after working at the workstation.

Never 4 3 2 Often

I have difficulty holding onto things.

Never 4 3 2 Often

I have trouble grasping items.

Never 4 3 2 Often

I wake up with neck pain after working at the workstation.

Never 4 3 2 Often

Mental Demand: Total: /40

I feel the workstation is safe.

Not safe 2 3 4 Safe

I am confident I will not be injured while at this workstation.

Injured 2 3 4 Not injured

I feel stressed about doing my job correctly.

Stressed 2 3 4 Not stressed

I feel frustrated with my duties at this workstation.

Frustrated 2 3 4 Not frustrated

I feel frustrated with my fellow workers at this workstation.

Frustrated 2 3 4 Not frustrated

I feel that the company takes safety seriously at this workstation.

Unconcerned 2 3 4 Serious

I feel that my fellow workers take safety seriously at this workstation.

Unconcerned 2 3 4 Serious

I take safety seriously at this workstation.

Unconcerned 2 3 4 Serious

Maintainability: Total: /25

This workstation is broken down often.

Often 2 3 4 Rarely

I must call a supervisor to the machine for problems often.

Often 2 3 4 Rarely

I trust this workstation to operate as it should.

No trust 2 3 4 Trust

When repair is necessary it is fixed quickly.

Too slowly   2   3   4   Quickly

I feel repairs are simple.

Don't know   Difficult   2   3   4   Simple

Likability: Total: / 50

I like working at this workstation.

Dislike   2   3   4   Like

Circle the things you like about the workstation and circle the rating, with 1 being low and 5 being high, to indicate how well you like each descriptor.

Design	1 2 3 4 5
Simplicity	1 2 3 4 5
Layout	1 2 3 4 5
Controls	1 2 3 4 5
Pace of work	1 2 3 4 5
Comfort	1 2 3 4 5
Low physical demand	1 2 3 4 5
Low stress	1 2 3 4 5
Reliable performance	1 2 3 4 5

Source: Dul and Weerdmeester 2008; Fitts and Posner 1979; Chapanis 1996.

In a usability assessment, safety performance is also a consideration. A human factors program manages where human workers meet technology. Since it is the role of the safety manager to advise the workstation designers and assessment teams as to the considerations of design, safety performance at the workstation becomes a consideration. For example, if a company was considering bidding on a contract to produce a certain part, it would usually bid on contracts that resemble current production capabilities evident by part similarity or capability of current production machinery or processes. The company might use its existing machinery and add as little new machinery and processes as required. Cost estimation would be based on projected production costs that rely mainly on current production costs and contracts for raw materials. Rather than adding safety in as an incidental cost of production, the safety manager should have real data on safety costs based on current company experience. Usability assessments by workstation are an ideal setting for recording such business-related data that reflect the usability of the workstation since much about the satisfaction of use is related to users' safety.

Throughout this safety management system, measuring success has been based on the Mathis structure of metrics that defines level 1 essentially as the duties of management under the program, the operational behaviors and perceptions as level 2, and performance indicators as level 3. The safety performance level of a usability assessment will serve as the basis for gathering performance data based on workstation. This will require the creation of a log or database for certain data items in reference to workstation and workstation characteristics. It is an extension of an injury log and may be combined with that log, if it is utilized. Other cross-functional departments may also have the need for similar databases. The maintenance department may create or use databases that can track criteria for assessing maintainability.

We should begin our log by workstation identification and certain safety descriptors. Besides specific identifiers, type of machinery can certainly categorize type in meaningful language for safety professionals. Power presses, cranes, grinders, woodworking machinery, welders, and laser equipment are all examples of machinery that by mere type stand out due to specific regulations covering those types of machines.

Protective devices are another example of safety characteristics that are important components to categorizing workstations. This category includes guards, devices, distance, location, and opening. Guards are enclosures that physically prevent people from reaching over, under, around, or through to a danger point. Safety devices prevent access during machine motion or stop machine motion during access but do not protect from a catastrophic failure. Examples include light curtains or pressure mats. Awareness barriers are another type of device, but these devices are not point of operation guarding. They alert personnel to the proximity of a danger point by audible or visual alarm. A common example is a back-up alarm on a forklift. Distance, location, and size of opening are types of positional guarding. Due to the position or opening, the danger point is out of normal contact range.

Along with protective devices is the type of control actuator, or control that enables or signals machine action. These are important safety descriptors because the type of actuator determines the correct safety device. Single-hand actuators, two-hand actuators, and foot pedals are typical examples. The important point is that the actuator and protective devices must combine to ensure the human user cannot contact a point of operation or danger point with any part of his or her body during machine motion.

From the satisfaction survey it is evident we are concerned with symptoms of workstation design inadequacies. When a human user is injured in any way, it can be viewed as a symptom of workstation design. This portion of the log is driven by the types of injury or symptoms of injury that may occur at a workstation. The other core programs play vital roles in workstation performance. The hazard recognition program is used to assess accurate participation, and the behavior-based safety program indicates operational dynamics. The box that follows is one possible example of such a workstation performance log.

WORKSTATION 011 WS 011

4/24/2018

General description: Parts produced:

Safety Descriptors

Classification of machine:

Protective devices:

Guards:

Point of operation device(s):

Awareness device(s):

Actuator(s):

Hazard Recognition

Participation accuracy: Level 1 hazards: _____

To-date totals:	Severity	Types		Symptoms	
Injury: ___	FA: ___	Cut: ___	Illness: ___	Hearing related:	Legs:
Illness: ___	Logged: ___	Abrasion: ___	Headache: ___	Sight related:	Back:
Condition: ___	Isolations :___	Avulsion: ___	Condition: ___	Forearm:	Neck:
		Amputation: ___		Arm:	
		Crush/pinch: ___		Wrist:	
		Burn: ___		Finger:	
				Shoulder:	
				Feet:	

Usability

Effectiveness: Y N	Efficiency error rate:	Satisfaction scores: ____ /360 Overall design: ____/50 Control interface: ____/ 35 Human comfort: ____/ 60 Physical demand: ____/ 75 Physical symptoms: ____/ 30 Mental demand: ____/ 40 Maintainability: ____/ 25 Likability: ____/ 50
Safety efficiency:		Total: ____/ 18 Lighting: Noise: Heat Physical risk:

Behavioral

Authorized average score:	Competent average score:	Gap score:

REFLECTION 8.1

1. What is the significance of using satisfaction surveying as an effort toward early identification?
2. How is human-centered design practiced in usability assessment?
3. What are failures and how can we classify them?

SYSTEM SAFETY ANALYSIS POLICY

System safety analysis policy is the glue that binds the organization together. It manages change through cross-functional division of skill and whole-team input. Whenever an organization assesses a workstation, machine, or process; puts into activity a new workstation, machine, or process; or makes a modification to an existing workstation, machine, or process, it should utilize its cross-functional talent to identify problems as early as possible and correct those problems.

Committee membership can be divided cross functionally, such as engineering, maintenance, safety, supervision, and workforce. The organization can have one or many standing committees organized efficiently in a manner that serves its needs. Committees can exist by lines of production, departments, or expertise.

The project is progressive and collaborative. Any progression toward completion requires sign-off approval and each cross-functional member has identified duties. There can be emergency procedures and checks in place to prevent bottlenecks from approval when immediate action is needed.

The policy will be structured with the following format:

- Purpose
- Scope
- Organization
- Activity and stages
- Division of duty

The box that follows reflects a possible policy.

SYSTEM SAFETY ANALYSIS

Purpose: The purpose of this policy is to utilize whole organizational talent from cross-functional teams in an effort to practice early intervention and continuous improvement in the launch of new workstations or modifications to workstations. Workstations include machines, processes, and positional assignments.

System safety analysis must be an organizational commitment to implement new machines or processes, and/or modify existing machines or processes in a systematic and team approach that practices early identification of risks and problems, solving them in early stages of implementation.

This will be accomplished through committee involvement that incorporates all levels of the organization. Committees will be organized by manufacturing line.

* Each line will form committees as needed for new machines or processes and modifications of existing machines or processes as needed. Assessments ordered after incident or from routine scheduling will be conducted by the human factors committee for the manufacturing line.

*Positional members of the committee are the following:

- Engineering department
- Maintenance department
- Safety department
- Line supervisor
- Line value stream manager

- Line associate
- Plant manager (may serve as signature/review member only serving to brief upper-level management)

The committee will report progress to the plant manager and management team as required on a regular basis throughout the implementation process.

Committee manager: The committee manager (CM) will serve as the chair of the committee.

Project type	CM
Launch of new process Launch of new machine Modification of a process Modifications of design, controls, PLC Modification of maintenance/inspection Quality control modifications	Engineering department
Human factor modifications of physical demand	Safety department
Modifications of LOTO process	Maintenance department
Work schedule modifications	Plant manager, department manager, line supervisor
Satisfaction reviews	Line associate

CM responsibilities:
1. Determine if _exigent circumstances_ exist
2. Schedule meetings and determine completion of stages
3. Notify and brief management or plant manager (when plant manager is acting as reviewer)
4. Assign committee roles of the following:
   a. Secretary
   b. Ethical over watch
   c. Devil's advocate
   d. Other roles as needed

New launch or non-exigent modifications
Stages of the process are as follows:

1. **Concept stage**: Presentation of need/idea/initial concepts are considered
2. **Prototype stage**: Initial design(s) are brought into being and are ready tested
3. **Initial testing stage**: Machine/process is initially trialed using engineering personnel
4. **Testing stage**: Test runs for production are initiated with line associates
5. **Initial production**: Production is completed with initial line associates on limited days/times

6. **Routine production**: The new process/machine is in a normal rotation of full scheduling, and production is possible with all line associates

General requirements:

1. The cross-functional department manager will determine and set duties for the departmental representative for each stage of analysis. These duties will be outlined in this policy as a general guide.

2. The *CM will not advance the project to the next stage of development until all cross-functional members have completed their duties and signed off on stage advancement.*

3. In the event a committee member is not available, the CM may replace the member permanently, assign a substitute, or conduct a meeting without the member, provided the member completes duties as required and reports results to the CM and briefs the committee prior to the next scheduled meeting.

Safety tasks for system safety analysis:

Concept stage	Research of like activity/machines Presentation to committee on findings/suggestions Review standard requirements Review past organizational experience with similar activity Initial safety cost analysis for bid/finance
Prototype stage	Identify hazard categories/sources (JHA) Audit/report on required safety procedures/standards/guarding/safety devices Produce ergonomic analysis Identify/produce safety training for line associate Review/document LOTO procedures
Initial testing	Complete observational review for ergonomic assessment/behavior-based assessments/training objectives/safety audits Train limited line associates
Testing	Review ergonomic/behavior-based assessments/training objectives with line associates Complete final safety audits Survey line associates for hazards/safety discrepancies
Initial production	Train all line associates for full production Release to full production
Routine production	Machine/process falls into the normal safety auditing/review rotation

Exigent circumstances:

Exigent circumstances are those situations of workstation modification that have an immediate need to counter a production or quality control issue to meet immediate shipment needs, advert an emergency, or respond to an emergency.

General requirements for exigent circumstances

- When exigent circumstances exist, engineering will determine the necessary duties and notifications of cross-functional team members required.
- Duties and notifications will be determined based on the immediate needs for safe conduct or to avert an emergency.
- All modifications made under exigent circumstances shall be referred to full committee review as soon as possible after the CM (assigned from engineering) determines the exigency of the situation to be over or as soon as possible during exigency.

Problem solving:

Safety incidents are incidents of near miss, property damage, or injury that involve infractions or possible infractions of OSHA regulations, adopted consensus standards, or company safety policy.

Methods of problem identification:

The safety department manager will assign lead investigator duties to safety department personnel and act as reviewer of the investigation.

Causal analysis will be formally conducted on all known incidents utilizing the following model:

Immediate cause level	Unsafe acts Substandard conditions
Basic cause level	Job factors Personal factors Human factors of workstation design
Root cause level	Management system factors

Incidents will be analytically reduced to identify incremental observable occurrences or events that comprise the whole incident.

Each event will be analyzed for cause.

Countermeasures will be developed based on the identification of causes for each incremental event to prevent further development of the incident.

Each critical incident will be mapped according to the butterfly methodology to show chronological occurrence, causes, and possible countermeasures.

Countermeasure will be developed to counter-causation based on the following hierarchy of controls.

Hierarchy of controls:

The development of countermeasures will utilize two concurrent strategies for formulating corrective actions to safety incidents.

Accident prevention: A psychological approach to safety relying on education and training initiatives	**Injury prevention**: An engineering approach relying on engineering controls aimed at eliminating, substituting, or reducing risk of injury from hazards, threats, and vulnerabilities

Policy/procedures
 	
Education
 	
Awareness of the following:
 	
    Sharing of organizational experience
 	
    Motivational postings
 	
    Safety/caution signage
 	
    Mailings
 	
    Away-from-work safety messages
 	
    Research findings
	1. Prevent the creation of the hazard.
	2. Reduce the amount of hazard brought into being.
	3. Prevent the release of the hazard that already exists.
	4. Modify the rate or spatial distribution of the release of the hazard from its source.
	5. Separate, in time or in space, the hazard and that which is to be protected.
	6. Separate hazards by barrier.
	7. Modify basic qualities of the hazard.
	8. Make that which is to be protected more resistant to the hazard.
	9. Counter the damage already done.
	10. Stabilize, repair, or rehabilitate the object of damage.

The policy maintains control of changes and utilizes cross-functional talent and skills. Policy can dictate formation and use of this committee, or a formulation of multiple committees, for specific concentration. Critical events can be reviewed for countermeasure approval or formulation, early symptoms of musculoskeletal disorders can trigger committee review for human factor design elements, and satisfaction surveys can facilitate assessment or trigger committee formation for countermeasure production. Committee practice and use follows the system safety analysis policy for team progression.

REFLECTION 8.2

1. What is the positive and negative potential for having accelerated procedures for modifications?
2. What is the significance between placing psychological safety controls as concurrent strategies to elimination, substitution, and engineering controls?

HUMAN FACTOR DESIGN ELEMENTS AND SAFETY EFFICIENCY

Human factors have evolved as a discipline because, as Chapanis (1996) says, "[G]ood design means not only designing for normal human use, but also designing against misuse, unintended use, and abuse" (10). Chapanis's standard for good design also meant design that considered human physiology, anthropometrics, and personal factors such as attitude, personalities, and characteristics. All core safety programs cumulate to allow the safety manager to efficiently allocate resources to the right problems at the right times to accomplish meeting the core duties of identifying hazards, threats, and vulnerabilities; assessing these problems; formulating corrective actions; implementing the corrections; and then improving the corrections in a cyclical manner. Hazard-recognition programs, behavior-based programs, and incident investigation programs are specific investigative and correctional programs that contribute to the human factors program as much as they cumulate to help deliver these main services to our primary customer, the workforce associate.

Work Agent to Controls Assessment

Controls are where work agents, or human and machine components, meet. There are five types of controls: manual, automated assist, voice activated, sensor activated, and manual computer control (Fitts and Posner 1979). Table 8.10 reveals the definition and examples for each type of control.

TABLE 8.10 *5 Types of Controls*

TYPE OF CONTROL	DEFINITION	EXAMPLE
Manual	A control that requires physical effort to transfer mechanical effort in direct correlation Tools can also be classified as manual.	Manual swing arms on derricks or hoists Steering wheels without steering assist Box-end wrench.
Automated assist	An assisted manual control that minimizes physical effort, amplifying applied effort Tools can also be classified in terms of automated assist when air, hydraulic, gears, etc. are used to minimize physical effort and increase applied effort.	Steering wheel with power steering Hydraulic swing arms Hydraulic blade controls on a dozer Pneumatic impact wrench

Voice activated	Action is initiated, stopped, or otherwise controlled by voice command keeping hands, feet, and eyes free for other tasks.	Hands-free smart phone controls on a vehicle
Sensor activated	Sensors activate or otherwise control action reducing the human role to monitor.	Automatic braking mechanisms Cruise control on a vehicle
Manual computer control	Computer controlled, or programmed action, from human input to a computer	Form of automated assist

Manual controls should require the least amount of physical effort as possible. This is why automated assist controls have dominated over manual controls. A good example is the blade control of a bulldozer, where the operator exerts minimal physical effort to signal hydraulic valves to place pressure on or move the blade. Voice-activated controls are preferred when the task divides attention and hands, feet, and eyes are needed for other tasks. Driving and simultaneous communication is a perfect example. It is important to note that hands-free communication while driving is more preferred than, say, texting and driving; cognitive load can impede paying attention. Sensor-activated controls reduce cognitive overload and move the human role more toward being a monitor. Continuing with the driving example, perhaps automatic braking features that utilize a sensor for gauging proximity engage when a person has failed to realize a dangerous scenario of a stopped frontal vehicle. Manual computer controls vary from traditional automated-assist controls in that the user is determining, to a minimal degree, when to signal or initiate movement. This is used when computer controls can coordinate multiple successive tasks to complete an operation. The human counterpart is deciding only when to initiate the whole operation and when to terminate it. A dozer operator must observe and feel the machine to manually signal changes to the blade. Computer-assist controls reduce the need for constant monitoring. A good example, continuing with the bulldozer example, is automated laser guidance controls performing finite movements and the operator engaging and disengaging the control depending on the dozer's position in relation to the work area.

The knowledge of control types and situational uses provides us with the first criteria for control assessment. Based on the human and machine interaction assessment we can recommend or assess for the preferred type of control for the situation that corresponds to the strengths and abilities of the human. Here, we mix human and machine interaction modeling to the types of controls for the situation. A good example is the human ability to monitor or measure. Let's examine a manufacturing work station where a formed piece of steel is to be drilled for a hole or a series of holes and then a nut is to be welded on the inside of the hole. To assure that the formed blank is placed in the exact position for precise hole placement, the controls must rely on sensors because this precise measuring capability is more the strength of a sensor/computer rather than from human measures. The process could begin with human placement of the blank on guides or with even more precise robotic arms placement. Either way, sensors will confirm the proper placement before allowing the next action to be performed. A user could then read sensor measurements and initiate more machine action, or it could be automated. As you see, the general rule is to reduce exposure of human weaknesses. Remember that exposure must be quantified as frequency, duration, and severity. In this case, it is exposure of human weaknesses and, according to human and machine interaction modeling, human weaknesses are the strengths of the computer or robot.

We wrestle with this concept because it is associated with replacing humans with machines. Workforce reduction is often a result or goal of lean manufacturing. But we can work to reduce loss by changing assigned tasks that result in reduced investments in higher technology and preservation of human positions based on cost projections. We can work to balance, in other words, rather than reduce human positions. Replacing humans with technology has a point at which it is more or less efficient. As a general guide, humans should be assigned to tasks high in the following:

- Sensory function (touch, hearing, smell, taste, sight recognition)
- Perceptual abilities (stimulus, conceptualization)
- Improvisation
- Judgment
- Selective recall
- Learning

Machines should be assigned to tasks high in the following:

- Alertness from sensors, readings, and programmed results
- Need for speed and power
- Sensor detection outside of human ranges
- Repetition
- Computation
- Short-term memory
- Simultaneous actions (Fitts and Posner 1979)

Technology is progressing with computers, and the line between strengths and weaknesses is getting more and more blurred. Since the late 70s, "deep thinking" has developed with computers and has allowed more and more features such as increased voice recognition and visual recognition. From the previous lists, tasks high in perceptual ability are human strengths while simultaneously programmable activities are computer strengths. Somewhere between these are divided attention tasks. Again, the concept is to aid humans in monitoring. Process-control room controls are good examples. Programming aids in alerting the human monitor to measures situations as they develop and can even suggest counters based on programmed experiences. However, the need for human ability to improvise and make judgment has not been eliminated. Table 8.11 can aid in assessing controls and tools to match with control type and work agent strengths.

TABLE 8.11 *Control Room Control Assessment*

CONTROL DESCRIPTION	CONTROL TYPE		TASK DESCRIPTION			
	Manual	☐	**Human**	**L**	**M**	**H**
	Auto assist	☐	Sensory function	☐	☐	☐
	Voice activated	☐	Perceptual abilities	☐	☐	☐
	Sensor activated	☐	Improvisation	☐	☐	☐
	Computer assist	☐	Judgment	☐	☐	☐
			Selective recall	☐	☐	☐
			Learning	☐	☐	☐
			Machine	**L**	**M**	**H**
			Alertness	☐	☐	☐
			Speed/power	☐	☐	☐
			Sensor detection	☐	☐	☐
			Repetition	☐	☐	☐
			Computation	☐	☐	☐
			Short-term memory	☐	☐	☐
			Simultaneous actions	☐	☐	☐

Assessing work agent strength to control type is an important follow-up to our initial assessment for assigning work tasks according to human and machine interaction, or work agent interaction. Rather than now working back to our initial assessment, we will include this work agent task assessment as our first component to safety efficiency in our usability assessment. In this concept we are rating the degree of reliance on agent strength. Table 8.12 reveals a matrix for this assessment.

TABLE 8.12 *Work Agent Task Assessment*

	Human low	Human medium	Human high
Machine high	Machine assigned	Machine assigned	Human assigned or machine assigned
Machine medium	Machine assigned/assist	Machine and/or human assigned with machine assist	Human assigned or machine assist
Machine low	Human assigned	Human assigned	Human assigned

As Table 8.12 reveals, the higher the task relies on human strengths, the more manual tasks or controls can rely on human strengths when control or tool interaction exists or is possible with a task. When initial assignment of this nature is made in considering usability, and once overall safety efficiency is evaluated, if the interaction falls outside of the acceptable range of the matrix, engineering can adjust or change the type of control and match design principals and feedback to the user.

Control design principals have developed from historical examination of human-centered design concepts since Chapanis first postulated that the work station's design can facilitate human error. The rules for proper design are as follows:

1. The control must anthropometrically fit the user.
2. The control's results must match user expectations.
3. Transfer effect must be reduced as much as possible
4. Placement of the control must be visible.
5. The control must have redundant coding.
6. The control must give timely feedback to the user.
7. The control must be compatible to the display.
8. Control functions should be limited.
9. Unintentional activation must be limited (Dul and Weerdmeester 2008; Fitts and Posner 1979).

The adherence to these rules accounts for control design efficiency as a component of safety efficiency in our workstation usability assessment. Table 8.13 suggests an assessment for design principle application to controls.

TABLE 8.13 *Workstation Usability Assessment*

CONTROL DESCRIPTION	CONTROL TYPE		CONTROL DESIGN DESCRIPTION			
	Manual	☐	**Design principle**	**L**	**M**	**H**
	Auto Assist	☐	Anthropometric fit	☐	☐	☐
	Voice Activated	☐	Match user expectation	☐	☐	☐
	Sensor Activated	☐	Transfer effect limited	☐	☐	☐
	Computer Assist	☐	Visible placement	☐	☐	☐
			Redundant coding	☐	☐	☐
			Timely feedback	☐	☐	☐
			Display compatibility	☐	☐	☐
			Functions limited	☐	☐	☐
			Accidental activation guards	☐	☐	☐

Agent interaction and control design evaluations serve as two components of workstation safety efficiency. At basic application of the concept we can rate the two criteria as low, medium, or high. We also must consider other environmental demands as a component to safety efficiency. Environmental demands are physical in nature, such as temperature, and they are mental, such as the organizational culture or relations between labor and management (Johnson and Maddox 2007). But in this individual assessment of a single workstation, we will limit assessment to physical nature of environmental demands, hoping to impact the mental nature holistically.

REFLECTION 8.3

1. Why are controls and feedback so important in relation to incident causation?

Environmental Demand and Safety Efficiency

It is important to cross reference the assessment of environmental demands conducted by safety and engineering, as a consultative matter with users, to the assessments made by the users in the satisfaction survey. As with all programs, in core and compliance, three levels of competency exist: authorized, competent, and administrative. The duties of competent-level personnel in the human factors program include conducting assessments of task effectiveness and task efficiency, ensuring completion of satisfaction surveys, and conducting safety efficiency surveys using consultation with existing users. The administrator must be able to compare satisfaction surveys with the other assessments and identify gaps.

Lighting

Lighting is an important physical demand of the environment due to its impact on the mental morale of the worker as well as having an impact due to physiology of sight. Dul and Weerdmeester (2008) suggest that for general tasks, the lux reading, which accounts for intensity and luminance, should be a minimum of 200 lux and a maximum of 750 lux. Specialized tasks, such as repairs, that might include fine measurements or finish grinding, require 750 lux up to 5,000 lux. Furthermore, they have suggested other rules for lighting at the workstation. These include the following:

- Avoiding excessive differences in lighting from one area to another
- Ensuring good legibility
- Screen sources of direct light
- Preventing reflections and shadows
- Combining ambient and localized lighting
- Avoiding flickering from fluorescent bulbs (Dul and Weerdmeester 2008)

Noise

Noise is another troublesome environmental concern. In assessing noise at a workstation, we can measure the overall decibels the worker will be exposed to with a dosimeter placed at the workstation. This will include the level of the workstation and all noise from other machines and processes that affect the station. We must, at minimum, meet OSHA requirements for noise dosage. But our overall goal is to reduce noise level to an exposure that minimizes error and maximizes user satisfaction. This might vary from situation to situation, and over time user satisfaction and incident tracking may reveal a level that is desirable. This is because noise is also feedback. For example, I like to be able to hear my engine when operating heavy equipment because it provides feedback as to load on the machine and mechanical problems.

Since noise is feedback, we might initially divide noise levels based on OSHA requirements and then adjust for a desired level that is lower than one hundred percent dosage. A highly efficient level may be a reading near 70 decibels. A recent study of worker performance in an automotive assembly setting found that productivity was high and efficient at the 70-decibel level (Akbari, Dehghan, Azmoon, and Forouhharmajd 2013). Initially we may categorize noise in determining safety efficiency as efficient between 75 decibels and 85 decibels. The rating of less efficient is a reading between 85 and 90 decibels. Of course, this is measured as a time-weighted average with hearing attenuation. As our ratings and program progresses, we can set a different lower level as a higher degree of efficiency and also begin to cross reference readings, either raw or with attenuation, based on the workstation setting. Front-office settings, for example, may need an ambient noise level as low as 50 decibels without hearing attenuation for that type of work. A 2005 study looked at low dose noise and human performance and found no difference with performance at 30-decibel background levels and 50 decibel-background levels. Thus, we can use an overall level of 50 as a target of efficiency for office- or clerical-type workstation tasks (Luszczyriska, Dudarewicz, Waszkowska, Szymczak, Kamedula, and Kowalska 2005).

Heat

Temperature is another factor of significant concern as an environmental and mental demand that should be considered in safety efficiency. Wet bulb globe temperature (WBGT) is a commonly used measure for establishing work and rest ratios for heat stress. It is applicable outdoors and indoors, provided the instrument used has those options or a different formula for calculation is used for indoor conditions. The difference in the two formulas is that when calculated for indoor conditions the calculation does not include adding the dry bulb or ambient temperature of the air. When used outdoors, the calculation is formulated from the following equation: 0.7 (natural wet bulb) + 0.2 (globe temperature) + 0.1 (dry bulb) = outdoor wet bulb globe temperature reading (Finucane 2006).

WBGT readings are applicable because heat stress arises from a variety of factors, primarily ambient temperature, relative humidity, the level of physical effort exerted by a task or job, and the clothing and gear worn by the person working in the environment. WBGT accounts for relative humidity and the effect on how the body cools itself through the process of sweating and evaporation. When evaporation of sweat is impeded, the ability for the body to cool is reduced (Finucane 2006). Clothing or gear can impede this process and places a higher physical load on the wearer. WBGT alone does not account for clothing or gear and physical effort. Therefore, we must cross-reference the WBGT reading with an assessment of physical effort and of clothing and gear. Clothing with lower air and vapor permeability and higher insulation values impede natural cooling (National Institute for Occupational Safety and Health, 2016). Physical effort of the job or task also must be evaluated, and it is important to note that acclimation to heat, physical fitness and health, and even age play a role. Sweat glands react slower in older individuals and they have reduced skin-level blood flow, which results in less effective cooling of the body (National Institute for Occupational Safety and Health 2016). National Institute for Occupational Safety and Health (NIOSH 2016) lists the risk factors for heat related illness as the following:

- High temperature
- High humidity
- Direct sun exposure
- Indoor radiant heat sources
- Stagnant air
- Dehydration
- Physical exertion
- Clothing
- Gear
- Physical conditioning
- Health problems
- Medications
- Pregnancy
- Lack of recent exposure
- Advanced age
- Suffering of a previous heat-related illness

Specific compliance programs for heat-stress countermeasures can and should be developed, especially for environments with higher WBGT conditions. NIOSH has made several recommendations for specific situations that can be referenced for those direct environments or for similar conditioned environments in its 2016 publication "Criteria for a Recommended Standard." In general, for human factor considerations and establishing safety efficiency during the design and modification processes, we can use the WBGT rating to establish a target for efficiency. NIOSH,

International Organization for Standardization (ISO), American Industrial Hygiene Association (AIHA), Occupational Safety and Health Administration (OSHA), and American Conference of Government Industrial Hygienists (ACGIH) all have standards or make recommendations based on WBGT readings and physical workload. Please note that OSHA has not promulgated a standard but has made recommendations since the early 1970s. We can use the NIOSH levels as a means for targeted efficiency; matching workload to WBGT ratings is considered efficient. Table 8.14 reflects the NIOSH levels.

TABLE 8.14   *NIOSH Levels*

PHYSICAL WORKLOAD	NIOSH WBGT LEVELS
Resting	33°C or 91.4 degrees F
Light	30°C or 86 degrees F
Moderate	28°C or 82.4 degrees F
Heavy	26°C or 78.8 degrees F
Very Heavy	25°C or 77 degrees F

WBGT readings are commonly measured in degrees Celsius or Fahrenheit. The ISO standard also includes the reading of 33 degrees C or 91.4 degrees F as a level of resting workload (NIOSH 2016). NIOSH now has recommended work and rest ratios per hour based on workload and WBGT. In this program we will rely on worker reports from the satisfaction survey for determining workload. However, there are ways to assess physical workload in a more objective and quantifiable manner. These would require medical monitoring and partnership. We will rate efficiency based on WBGT reading and workload rated by worker survey. Work and rest ratios can be followed administratively as conditions vary on workdays or as they change with the seasons. We will note the date and general weather conditions during assessment. Outdoor conditions do influence the indoor conditions in most manufacturing settings. Therefore, Table 8.13 represents efficient heat-related conditions. Less efficient ratings would apply to situations where the workload is outside the range of ideal WBGT conditions and would therefore require a work rest ratio that fits the situation. NIOSH has published several work-rest ratio tables that apply to various situations, such as work in chemical protective suits and work in typical clothing. The administrator of the program must adopt and train personnel on the appropriate work/rest tables that apply to situation.

Cold stress is another concern and usually involves decreased circulation of skin-level blood and extremities. Exposure to cold temperatures and vibration are primary causes (Finucane 2006). Much mental stress accompanies work tasks that involve frequently changing from cold to heat. Regardless, administrative controls and protective clothing comprise the majority of abatement strategies. Cold stress will be assessed in the satisfaction survey.

Motion, Posture, and Lifting

Throughout most of my life and career I have lived in weight rooms training hard for sports, military, or policing, but mainly to make myself a better person through trial. When I became responsible for safety in such a facility, it occurred to me how important behaviors were to safety. Advanced lifters would practice perfect technique during their lifts but rarely lift properly while they were retrieving or replacing weights from the storage trees. It used to be common to place the heavier plates near the bottom of the weight tree. This only compounded the problem. Lifters would routinely bend their back with a stiff leg to lean down to grab or replace the 45-pound plates. To counter this, I began placing the 45-pounders at the top to minimize bending over at the waist and performing a stiff-legged lift. Placing them at the top minimized the effects of improper lifting. It took a while for trainees to catch on, but now it is not uncommon to see such practice. The point is that we can spend lots of time training and educating on proper lifting, posture, and movement, but what is more reliable is to design a workstation to minimize the amount of conscious effort it takes to lift properly. This means developing known or found risk factors into assessable criteria.

Selecting or developing factors into criteria depends on the work station. For example, will the station require standing or sitting or both? Other considerations might be the amount of eye involvement and hand involvement and the amount of lifting, carrying, or push versus pull activity. The result of ergonomic risk factors associated with movement, posture, and lifting are primarily musculoskeletal disorders such as strains, muscular tears, ruptured discs, bulging discs, trigger finger, or carpal tunnel disorder. General risk factors in a manufacturing setting are as follows:

- Exerting high levels of force
- Repetitive tasks
- Awkward positions or postures
- Maintaining static body positions for extended periods
- Physical contact to an edged surface
- Vibration (US Department of Labor 2012)

From the general risk factors we can structure a basic set of criteria for initial assessment of safety efficiency in regard to physical risk factors. Table 8.15 suggests one such format for scoring physical risk factors and scoring this portion of human factors.

TABLE 8.15 *Risk Factor Scoring*

PHYSICAL RISK FACTOR	EFFICIENCY CRITERIA	SCORE	1	0
Exerting high levels of force	Manual lifting has been limited	Y ☐	N ☐	N/A
	Loads stay in optimal lift zone (Shoulders to waist/11.8 inches from spine)	Y ☐	N ☐	N/A
	Loads limited to less than 50 pounds	Y ☐	N ☐	N/A
	Loads limited to less than 26 pounds	Y ☐	N ☐	N/A
	Loads remain close to body	Y ☐	N ☐	N/A
	Adequate lifting space	Y ☐	N ☐	N/A
	Lift assists utilized	Y ☐	N ☐	N/A
	Loads have handles/lift grip	Y ☐	N ☐	N/A
	Loads shaped for lifting	Y ☐	N ☐	N/A
	Sudden force avoided	Y ☐	N ☐	N/A
Repetitive tasks	Excessive reaching avoided	Y ☐	N ☐	N/A
	Repetitive tasks limited	Y ☐	N ☐	N/A
	Muscle exhaustion avoided	Y ☐	N ☐	N/A
Awkward positions or postures	Joints remain in neutral positions	Y ☐	N ☐	N/A
	Bending over postures avoided	Y ☐	N ☐	N/A
	Twisting limited to less than 30 degrees	Y ☐	N ☐	N/A
	Work is within optimal work zone	Y ☐	N ☐	N/A
	Work is below shoulder height	Y ☐	N ☐	N/A
Maintaining static body positions for extended periods	Standing/sitting alternated	Y ☐	N ☐	N/A
	Worker changes position adequately	Y ☐	N ☐	N/A
	Rests after heavy work activity	Y ☐	N ☐	N/A
Physical contact to an edged surface	Excessive contact avoided	Y ☐	N ☐	N/A
Vibration	Jolts prevented	Y ☐	N ☐	N/A
	Transfer vibration limited	Y ☐	N ☐	N/A

Source: Dul and Weerdmeester 2008; Marras 2008; NIOSH 2016.

Pulling together the measurable components of safety efficiency (lighting, noise, heat, motion, posture, and lifting) into a quantified score is revealed in the box that follows.

Workstation Safety Efficiency

Component	Criteria	Efficiency rating	Total score
Lighting	General tasks 200–750 lux	Efficient (3) ☐ Less efficient (1) ☐ Inefficient (0) ☐	
Lighting	Specialized tasks 750–5000 lux	Efficient (3) ☐ Less efficient (1) ☐ Inefficient (0) ☐	
Noise (calculating work) (Non-attenuated reading)	50 decibels or less (efficient) 50–70 decibels (less efficient) 70 decibels and above (inefficient)	Efficient (3) ☐ Less efficient (1) ☐ Inefficient (0) ☐	
Noise (production work) (With attenuation)	70–80 decibels (efficient) 80–90 decibels (less efficient) Outside of TWA standard (inefficient)	Efficient (3) ☐ Less efficient (1) ☐ Inefficient (0) ☐	
Heat stress	Physical workload<table><tr><th>NIOSH</th><th>WBGT levels</th></tr><tr><td>Resting</td><td>33 degrees C or 91.4 degrees F</td></tr><tr><td>Light</td><td>30 degrees C or 86 degrees F</td></tr><tr><td>Moderate</td><td>28 degrees C or 82.4 degrees F</td></tr><tr><td>Heavy</td><td>26 degrees C or 78.8 degrees F</td></tr><tr><td>Very heavy</td><td>25 degrees C or 77 degrees F</td></tr></table>	Efficient (3) ☐	
Heat stress	Falls outside efficient rating for workload but adheres to situational work/rest ratio	Less efficient (1) ☐	

Physical risk factor	Efficiency criteria Score	1	0
Exerting high levels of force	Manual lifting has been limited	Y ☐ N ☐ N/A	
	Loads stay in optimal lift zone	Y ☐ N ☐ N/A	
	(Shoulders to waist/11.8 inches from spine)		
	Loads limited to less than 50 pounds	Y ☐ N ☐ N/A	
	Loads limited to less than 26 pounds	Y ☐ N ☐ N/A	
	Loads remain close to body	Y ☐ N ☐ N/A	
	Station has adequate lifting space	Y ☐ N ☐ N/A	
	Lift assists utilized	Y ☐ N ☐ N/A	
	Loads have handles/lift grip	Y ☐ N ☐ N/A	
	Loads shaped for lifting	Y ☐ N ☐ N/A	
	Sudden force avoided	Y ☐ N ☐ N/A	
Repetitive tasks	Excessive reaching avoided	Y ☐ N ☐ N/A	
	Repetitive tasks limited	Y ☐ N ☐ N/A	
	Muscle exhaustion avoided	Y ☐ N ☐ N/A	
Awkward positions or postures	Joints remain in neutral positions	Y ☐ N ☐ N/A	
	Bending over postures avoided	Y ☐ N ☐ N/A	
	Twisting limited to less than 30 degrees	Y ☐ N ☐ N/A	
	Work is within optimal work zone	Y ☐ N ☐ N/A	
	Work is below shoulder height	Y ☐ N ☐ N/A	
Maintaining static body positions for extended periods	Standing/Sitting alternated	Y ☐ N ☐ N/A	
	Worker changes position adequately	Y ☐ N ☐ N/A	
	Rests after heavy work activity	Y ☐ N ☐ N/A	
Physical contact to an edged surface	Excessive contact avoided	Y ☐ N ☐ N/A	
Vibration	Jolts prevented	Y ☐ N ☐ N/A	
	Transfer vibration limited	Y ☐ N ☐ N/A	
Total physical risk score: Possible score:	Efficient (3) ☐ 90 percent and above		
	Less efficient (1) ☐ 80–89 percent		
	Inefficient (0) ☐ Below 79 percent		

Total safety efficiency rating: / 18

In this assessment form, physical risk factors are figured into an overall rating, making the total for safety efficiency a possible 18. This form can be modified to include more criteria as the situation demands.

REFLECTION 8.4

1. Does the concept of safety efficiency parallel the concept of reducing waste in lean practices? Why or why not?

HUMAN FACTORS REVIEW FOR EFFECTIVENESS

Like any core or compliance initiative, continual improvement demands annual review of the human factors program as well. Tables 8.16–18 shows some possible program effectiveness metrics for human factors.

TABLE 8.16 *Management-Level Criteria*

ITEM	CULTURE CRITERIA	CRITERIA	RESULT
1	Participation	The organization has developed and implemented a system safety analysis policy. The following components constitute scoring elements: • Cross-functional team review • Stages of development • Early identification of problems • Abatement strategy • Emergency procedures	1 2 3 4 5
2	Participation	The organization has upper-level management assigned to human factor committee review.	1 2 3 4 5
3	Commitment	The organization has approved resource requests from committee recommendations adequate to reduce risk to an acceptable level.	1 2 3 4 5
4	Commitment	The organization provides training to associates at a level that allows adequate participation.	1 2 3 4 5

Continued

ITEM	CULTURE CRITERIA	CRITERIA	RESULT
5	Competency	Committees have management-level personnel developed to at least the competent level for all committees or for those assigned to sign off on committee reports.	1 2 3 4 5
6	Compliance	The company adheres to the system safety analysis policy.	1 2 3 4 5

TABLE 8.17 *Cultural and Operational-Level Criteria*

ITEM	CULTURE CRITERIA	CRITERIA	RESULT
1	Perception	Workforce associates perceive the program to be important.	1 2 3 4 5
2	Perception	Workforce associates perceive management as viewing the program as important.	1 2 3 4 5
3	Perception	Workforce associates perceive the company as following policy.	1 2 3 4 5
4	Perception	Management-level associates perceive the program to be important.	1 2 3 4 5
5	Perception	Management-level associates perceive the company as following policy.	1 2 3 4 5

TABLE 8.18 *Status and Performance-Level Criteria*

ITEM	CULTURE CRITERIA	CRITERIA	RESULT
1	Performance	Percentage of failures eliminated during review process: _____	1 2 3 4 5
2	Performance	Number of hazards found within sixty days of release of workstation to routine production status _____	1 2 3 4 5

3	Performance	Number of failures within sixty days of release of workstation to routine production status: _____	1 2 3 4 5
4	Performance	Average satisfaction score for workstations: _____	1 2 3 4 5
5	Performance	Average safety efficiency score for workstations: _____	1 2 3 4 5
6	Performance	Sixty-day BBS composite score: _____	1 2 3 4 5

The administrator of the initiative can determine scoring based on performance of the organization as whole, for metrics such as sixty-day BBS composite scores or set standards, such as zero hazards within the first sixty days as correlating to a 5 score. Perception surveys can parallel scoring based on the average rating by associate levels. But review of initiative effectiveness is a core concept for continual improvement.

REFLECTION 8.5

1. Why are each of the criteria important for program review?
2. Why are the criteria assigned as a measure for culture?

CONCLUSION

The human factors program concentrates on human and machine interaction for assessment of effectiveness and efficiency, user satisfaction, and safety efficiency as completing usability assessments of production lines, equipment, processes, and machinery. It is proactive in that it practices early intervention in new-launch projects or modification projects. System safety analysis policy becomes a glue that ensures cross-functional team input and review for this process. The program is central in the management of musculoskeletal disorders in that anytime a symptom is reported cross-functional team review of the workstation and task is ordered. Without a human factors initiative, continual improvement for design factors is absent, resulting in repeat incidents. Just as compliance-level auditing is central for proactive uncovering of substandard conditions and behavior-based safety is for unsafe acts, a human factors program assesses job factors and workstation design factors that can facilitate error and result in loss.

REFERENCES

Akbari, Jafar, Habibollah Dehghan, Hiva Azmoon, and Farhad Forouharmajd. 2013. "Relationship between Lighting and Noise Levels and Productivity of the Occupants in Automotive Assembly Industry." *Journal of Environmental and Public Health*, no. 52707.

American National Standards Institute. 2008. *B11-2008 General Safety Requirements Common to ANSI Machines*. McLean, VA: Association for Manufacturing Technology.

Bird Frank E., Jr., George L. Germain, and Douglas M. Clark. 2003. *Practical Loss Control Leadership*, 3rd ed. Duluth, GA: Det Norske Veritas.

Boy, Guy, A. 2011. "A Human-Centered Design Approach." In The Handbook of Human-Machine Interaction: A Human-Centered Design Approach edited by Guy Boy, 1–20. Farnham, UK: Ashgate.

Chapanis, Alphonse. 1996. *Human Factors in Systems Engineering*. New York: John Wiley and Sons.

Dotson, Ron. 2017. "Human Factors for Safety Managers." In *Principles of Occupational Safety Management*, edited by Ron Dotson, Troy Rawlins, Earl Blair, and Scott Rockwell, 125–144. San Diego, CA: Cognella.

Dul, Jan, and Bernard Weerdmeester. 2008. *Ergonomics for Beginners: A Quick Reference Guide*, 3rd ed. Boca Raton, FL: CRC Press.

Finucane, Edward, W. 2006. *Definitions, Conversions, and Calculations for Occupational Safety and Health Professionals*, 3rd ed. Boca Raton, FL: CRC Press.

Fitts, Paul, M. and Michael Posner. 1979. *Human Performance*. [AU: Add city of publication]: Greenwood Press.

Johnson, William, B. and Michael E. Maddox. 2007. "A Model to Explain Human Factors in Aviation Maintenance." *Avionics News*, April, 38–41.

Luszczyriska, Malgorzata, Adam Dudarewicz, Malgorzata Waszkowska, Wieslaw Szymczak, Maria Kamedula, and Mariola Kowalska. 2005. "Does Low Frequency Noise at Moderate Levels Influence Human Mental Performance?" *Journal of Low Frequency Noise, Vibration, and Active Control* vol 1, no. 1, 25–42.

Marras, William, S. 2008. *The Working Back: A SystemsVview*. Hoboken, NJ. John Wiley & Sons.

Meister, David. 1971. *Human Factors: Theory and Practice*. New York: John Wiley and Sons.

National Institute for Occupational Safety and Health. 2016. *NIOSH Criteria for a Recommended Standard: Occupational Exposure to Heat and Hot Environments*. Washington DC: US Department of Health and Human Services.

U.S. Department of Labor, Occupational Safety and Health Administration. 2012. *Solutions for the Prevention of Musculoskeletal Injuries in Foundries* (OSHA Publication No. 3465-08). https://www.osha.gov/Publications/osha3465.pdf

Figure Credits

CHAPTER NINE

Structuring the Program to Meet ISO 45001

FOREWORD

Safety management systems give guidance on overall program structure to help organizations have comprehensive systems that enable culture. The new ISO 45001 International Standard that was recently published in March of 2018 provides general guidance for system dynamics. Requirements such as continuous improvement efforts, management review, hazard recognition, and hazard abatement are described to ensure best management practices are in place.

This chapter will explore the general structure of compliance programs and will also examine the requirements of ISO 45001 and how the management system described throughout this text meets these guided requirements.

Objectives

After reading this chapter the learner will be able to do the following:

1. Identify the difference between compliance initiatives and core management initiatives
2. Identify typical compliance and core management initiatives
3. Outline the structure of compliance initiatives
4. Formulate compliance initiative metrics
5. Identify the guided requirements of meeting ISO 45001
6. Assess the management system formulated in this text for meeting ISO 45001 requirements

LEARNING PLAN

LEVEL	ULTIMATE OUTCOME	CLAIM	LEARNING TYPE	ASSESSMENT
F	Identify the difference between compliance initiatives and core management initiatives. Identify typical compliance and core management initiatives.	Core and compliance initiatives serve different goals.	CR/AC	Student will be able to recognize or list the differences between core- and compliance-level management initiatives.
U	Formulate compliance initiative metrics.	Measuring program effectiveness is a function of management review and a critical component for continual improvement.	CT/AC	Using the structure for program metrics provided in chapter 2, students will be able to formulate compliance program metrics.
U	Outline the structure of compliance initiatives.	Compliance initiative structure is necessary for accurate policy application.	CT	Student will create a draft compliance initiative with possible or fictitious information.
U	Identify the guided requirements of meeting ISO 45001. Assess the management system formulated in this text for meeting ISO 45001 requirements.	Assessing safety management systems for compliance to standards is an important skill for system management.	CT	Students will develop a compliance-level audit using the covered data from ISO 45001 for formulation of audit criteria to assess a safety management system for meeting basic ISO 45001 requirements.

Learning Plan Legend
    Level:
    F: Foundational outcomes: Basic abilities
    M: Mediating outcomes: Progress through a developmental model; interpret, analyze, evaluate progressively challenging claims, arguments
    U: Ultimate outcome: Navigate most advanced arguments/claims

Type of Learning:
CR: <u>Critical reading</u>: The ability to read, process, and understand the meaning of written information
IL: <u>Information literacy</u>: Locating and selecting suitable information for a task; evaluating appropriateness/validity of information sources
AC: <u>Application of concepts</u>: Ability to apply discipline-specific knowledge/skill to tasks/situations important to the discipline
CT: <u>Critical thinking</u>: Ability to apply a concept to a vague or argumentative claim without a creative leap
AT: <u>Analytical thinking</u>: Ability to critique/analyze situations using a concept or model
CA: <u>Creative application</u>: Ability to apply a model/concept in a new way/to an unrelated situation or scenario; involves creative leaps

COMPLIANCE INITIATIVES

This entire management system begins with the identification of customers and their needs and devising core management programs that allow the safety department to fulfill its core duties. These core programs allow for the management of safety, or the act of allocating resources at the right problems in a timely efficient manner. This is how the scope of the system is set.

Compliance programs are specific to activity, situation, hazard, threat, or vulnerability. A regulation or standard usually exists that provides guidance for procedures to perform or abate the hazards in a safe manner. The focus is on complying with legal requirements for safety. Examples of compliance programs include lock-out tag-out, forklift operations safety, machine guarding, fall protection and prevention, walking and working surfaces, hazard communications, emergency response and fire prevention, confined spaces, electrical safety, and various exposure programs such as blood-borne pathogens. The hazard inventory also assesses the legal liability for the facility and provides a general view of what specific compliance programs will be needed to meet the liability.

Compliance programs are policy. But they are also reference guides to be used to safely conduct activity-abating specific hazards. Design of the program should facilitate it being used as a reference guide by relying more on "how to" rather than a collection of policy demands. For example, rather than a policy demand requiring a worker to wear fall protection when working at a height above a certain level, a program designed as a reference guide would also include schematics and procedures for appropriate protection in situations typically encountered. It is for this reason compliance programs are constantly updated with specific situations or new techniques are encountered by the organization. Figure 9.1 reveals a possible template for safety policy. Compare it to the template for a safety initiative that will be covered next.

Safety Policy Template Form

TITLE OF POLICY

Policy area:	**Policy number:**
Effective date:	**Revision date:**
Approved by:	

    Purpose:

    Reference:

    Scope:

    Definitions:

    Policy statement:

    Policy review/measures:

    Advising:

FIGURE 9.1   Safety Policy Template

The purpose is a goal that can be statistically measured, observed, or demonstrated. References are sources for which the policy is based and are important for sustainability as they tell future reviewers why the policy is as it is. Scope is the limit of coverage, which might be geographical as well as situational. Definitions are for the key words or phrases to make words with vague or multiple meanings applicable to the policy. The policy statement includes general requirements and may also include roles and responsibilities. Policy review and measures are where specific measures of compliance, review dates, and methods are covered. For example, the policy may be reviewed by a manager-level committee every three years. Advisors are experts on the policy who can be accessed for more information.

STRUCTURING COMPLIANCE INITIATIVES

To distinguish between the overall safety program and compliance programs, compliance programs can be called initiatives, core management programs can also be called core initiatives, and finally, program will, from this point forward, be called the safety program.

Compliance program structure begins with a title page. The title page contains the title or name of the overall safety program, symbolism, the title or subject of the specific compliance program, and document control information in a header and/or footer. Symbolism such as a slogan or picture is helpful in communicating to members within the organization. Document control data includes the version of the program, any unique control number assigned by the organization, the person who has overall responsibility for the program, when it was approved, and possibly a future review date. Document control data is contained on every page of an organizational document to prevent varying versions from being used within the organization. When a new version of the document is developed, the old documents are retrieved, destroyed, and replaced with the current version. The compliance program structure also may communicate the privacy level of the program, such as confidential or for executive management only. Most safety compliance programs are open to all internal members of the organization; however, some may contain sensitive information, or more advanced versions of the same program may exist. A good example of this exists with hazardous materials and the protection from sabotage or theft of certain substances. Figure 9.2 shows a possible cover page.

The next page of the compliance program is a log of issue and log of significant changes. This reveals the history of the program and documents who has copies or where copies have been issued to retrieve older versions for replacement. Figure 9.3 reveals a program history page.

The management page contains information about the purpose of the program and reveals or states overall goals. It communicates the scope of the program, to include what facility and operations it is applicable to. It is specific as to availability of the program and also contains references. References of the origins of the contents are important for management review and program update, as well as sustainability of the program. References will list the regulations, national consensus standards, state laws, and other sources of information that help develop the program. If a copy of these are included, these should be included as an appendix. If the initiative has a committee for review or for problem solving, the committee should also be identified on the management page. Figure 9.4 is an example of a template for a management page of a compliance program.

The central elements of the program begin with defining key or unique terms to the program. It continues with the following list of elements: responsibilities, general requirements and procedures, training, equipment specifics, auditing, and any appendices. The auditing section will contain the program effectiveness level audit, which reveals the metrics to be tracked for the initiative. Figure 9.5 is a template that serves as a general guide for program structuring beyond the management pages.

XYZ CORP   Lock-Out / Tag-Out Initiative
INTERNAL ASSOCIATES ONLY
XYZ CORP SYMBOL

Title of Safety Program
LOTO

DATE Issued:
VERSION: 1.0
DATE of Next Review:

Version 1.0   Date
Approved By: Safety Manager   Page 1 of x   Company control #

FIGURE 9.2   Cover Page of a Compliance Program

XYZ CORP   Lock-Out / Tag-Out Initiative

INTERNAL ASSOCIATES ONLY

LOTO INITIATIVE HISTORY

ISSUE LOG

Version	Department/Associate	Received By:	Date

Revise Date	Revision	Signature

Version 1.0   Date
Approved By: Safety Manager   Page x of xx   Company control #

FIGURE 9.3   Program History Page of a Compliance Program

XYZ CORP   Lock-Out / Tag-Out Initiative

INTERNAL ASSOCIATES ONLY

Management Page

Program Goal:

Goals	Performance Measure

Specific performance measures should come from the Performance or Status (Level III) Metrics of the effectiveness audit.

Availability:

A copy of this program will be kept in the Safety Office, Human Resource Office, and all Departmental Offices. Copies will be made available to any associate upon request.

	Room #	Phone #	Contact Associate
Human Resource:			
Press Dept.:			
Weld Dept.:			
Shipping/Receiving:			
Maintenance:			
Engineering:			

Scope:

This initiative will apply to all facilities owned and operated by XYZ Corp.

Plant 1   Plant 2   Plant 3
Address   Address   Address
Contact   Contact   Contact

- May pertain only to certain operations/employees/processes. The policy statement should list specific application. This may come from compliance standards and company job descriptions.

Standing Committee:

Department	Title/Position	Name

Version 1.0   Date
Approved By: Safety Manager   **Page x of xx**   **Company control #**

XYZ CORP   Lock-Out / Tag-Out Initiative

INTERNAL ASSOCIATES ONLY

Reference Page

This page includes a list of all referenced standards to include consensus standards.

Referenced Standards:

29 CFR 1910.147

29 CFR Subpart S

NFPA 70E

After these are listed, you should include where these are kept or whether a copy is attached to the program as an appendix.

Definitions:

This section has two strategies of formulation. The first strategy is to list definitions that have unusual, unique, or specific meanings to the policy or company. This includes terms that are skill or trade specific and may have more than one interpretation. The second strategy is to copy all of the definitions from the referenced OSHA standard and consensus standard. Copying the definitions from a standard may not be necessary if you attach the standard to the initiative as an appendix as it becomes too redundant and makes the document too lengthy. When a document is too lengthy or complex, it loses its appeal to field associates as a reference guide. Initiatives should be written in as reader friendly and appealing manner as possible. This means communicating data in visual methods and limiting paragraph style writing by listing, bullet pointing, and including visuals.

Version 1.0   Date
Approved By: Safety Manager   *Page x of xx*   *Company control #*

FIGURE 9.4   Management Page of a Compliance Program

XYZ CORP   **Lock-Out / Tag-Out Initiative**

INTERNAL ASSOCIATES ONLY

TABLE OF CONTENTS

Version 1.0   Date
Approved By: Safety Manager   *Page x of xx*   *Company control #*

XYZ CORP   **Lock-Out / Tag-Out Initiative**

INTERNAL ASSOCIATES ONLY

Responsibilities

The responsibilities under any initiative are the first section meant as a quick reference guide. It quickly shows the associate competency requirements and job duties. Building upon the basic structure of initiatives described in the Employee Oriented Safety System, this is divided as authorized, competent, and administrator. It may helpful to also list the job title or names of personnel that fit or fill these levels. As the program matures, this will change to include more workforce level associates at higher levels of competency.

Level	Personnel	Competency	Duties
Authorized	All associates	Training: Experience:	
Competent	Supervisors of Line Team Leads	Training: Experience:	
Administrator	Safety Manager	Training: Experience:	

Responsibilities may be further explained as needed.

XYZ CORP   Lock-Out / Tag-Out Initiative

INTERNAL ASSOCIATES ONLY

General Requirements:

This is the heart of the initiative and contains many policy statements. It tells all associates what to do to meet compliance with the initiative. Again the difference between policies and initiatives is that an initiative is meant to be a "how to guide" for meeting safety compliance. This section should include summarization tools like charts, graphs, pictures, maps, etc. that explain and enhance any narrative explanation. If forms are required to document action then forms should be included, either at this point or as an appendix. Flow charts that show chain of communication or sequential steps is another helpful tool for guidance.

Since humans do not like to read paragraphs and pull out specific information, using bullet points, short sentences, plain and simple language, and having good organization is required in order to make an initiative a reference guide.

Any educational posting associated with the initiative should also have an official tracking number on it. This number ensures that all postings are official and that old versions will not be on the floor when a new version has been issued. When a standard posting is used, location of the postings should be covered. References to postings may be necessary in a general requirements section.

Organization and structure of the general requirements section can vary. It can be by activity, level of competency, or situational. But it is sequential in nature. It may resemble the example below. Compliance initiative policy relies heavily on a standard for development. This example begins at 1910.147(a)(1)(i) and clearly communicates what situation the policy is applicable to, but also gives guidance for policy covering similar or related situations.

   1.0 Applicable Situations
      1.1 This initiative covers all _servicing and maintenance_ of machines or equipment where the _unexpected startup or release of stored energy_ could cause injury to any associate.
      1.2 Situations requiring troubleshooting must refer to the _Live Work Section of the Electrical Safety Initiative._

Version 1.0 Date
Approved By: Safety Manager   _Page x of xx_   _Company control #_

XYZ CORP   **Lock-Out / Tag-Out Initiative**

INTERNAL ASSOCIATES ONLY

Training:

The training section also has guidance from standards, but must be more specific to the organization. Generally, the section must cover who is to be trained, who can perform the training, what other or outside resource training will be used or accepted, when refresher training is to be conducted, and how training will be documented and tracked. This may be divided and structured similarly to the responsibilities section.

Lesson plans are also important components of the training section. Lesson plans in whatever format they are developed and structured, to include direct copies of power points serve two important functions for safety management. It serves as a basis for assessing training effectiveness and it serves as a justification for any disciplinary action. Foundational principles of justice tell us that discipline must be appropriate to the level of established competency.

XYZ CORP   **Lock-Out / Tag-Out Initiative**

INTERNAL ASSOCIATES ONLY

Equipment Requirements:

Many organizations cover equipment issues in training content. But it is important to have established policy on equipment standards. This includes what equipment specifications or features are required, acceptable brands or types, inspection for serviceability, and how to report issues and properly dispose of equipment. Specific to electrical safety may be specific requirements to insulated tools and how to recognize a condition that renders a tool unserviceable. Pictures become vital.

XYZ CORP   Lock-Out / Tag-Out Initiative

INTERNAL ASSOCIATES ONLY

Auditing:

In the Employee Oriented Safety System audits are contained within each initiative. There are two levels of initiatives; compliance level and effectiveness levels. Auditing is an important investigational technique and every initiative must rely on compliance and continual improvement through management review.

XYZ CORP   Lock-Out / Tag-Out Initiative

INTERNAL ASSOCIATES ONLY

Appendices:

The appendices section will include all mentioned organizational forms and documents, postings, and it may include referenced standards. But do not overwhelm potential users with a lengthy document. Perhaps keeping all desired appendices in the master or safety copy of the initiative, and issuing a scaled down version for field use is desirable.

Version 1.0   Date
Approved By: Safety Manager   Page x of xx   Company control #

FIGURE 9.5

MEETING ISO 45001

The International Organization for Standardization (ISO) has published the latest safety management system standard with the intended outcomes of helping an organization practice continual improvement in safety performance, meet legal and other requirements associated with its operations, and achieve its safety goals (ISO 45001 2018). It is widely accepted that establishing a system of management that meets organizational needs helps achieve those goals and sustain achievement through business interruption and personnel turnover.

Foundational Comparison of Workforce Involvement and Effectiveness

The safety management system described and established in this text meets this standard and pushes it to the next level in many foundational manners. The first concept that is substantially different is employee involvement. The ISO standard identifies two levels of involvement: consultation and participation. The concepts are not defined differently in employee-oriented safety, but rather include and have the goal of the next level, which is empowerment. Involvement begins at the consultation level where workforce-level associates have input prior to decision making. Participation then is the next step of allowing workforce-level associates to be involved in decision making itself, usually by committee or group involvement and equal voting power. But the ultimate level of participation is empowerment. Empowerment is where workforce-level associates have authority to make decisions and take complete action or initiative.

Empowerment in core management programs reflects safety as becoming a virtue or value in all decision making that does not yield based on the situation. Empowerment cannot exist without development toward self-actualization. As the criteria of culture develops, more mature programs begin to take on the picture of having workforce-level associates assume competent-level responsibilities and in some cases take over administrator duties for individual initiatives. This structuring philosophy allows safety professionals to perform overall administrator duties for several initiatives, with the safety manager having overall administrator duties for the program. This is especially functional because a safety department typically has fewer personnel as compared to other cross-functional departments.

Effectiveness is an important component to continual improvement and accountability, both core values of ISO 45001. In ISO 45001, effectiveness is defined as the "extent to which planned activities are realized and planned results achieved" (Clause 3.13, page 4). Employee-oriented safety (EOS) builds effectiveness into all initiatives and takes the concept a step further in reviewing initiatives by examining the degree of which initiatives are implemented or realized at the management level and operational level in comparison to final performance measures or planned results. EOS also examines effectiveness of countermeasures to incidents and proactive compliance audits.

Meeting the Requirements of 45001

Requirements for ISO 45001 begin with examining the context of the organization. This means establishing the scope of the safety management system in light of operations and needs. EOS accomplishes this with the identification of customers and core initiatives to provide services that meet customer needs. Customers are not merely internal customers of the organization but of the surrounding community, which are the "other interested parties" ISO 45001 requires to be included in its scope.

Context or overall scope of the system must also identify the legal and other liabilities that the organization's operations present, both internally and external to the organization, and have initiatives that cover these liabilities (ISO 45001 2018, Clause 4). The inventory concepts of EOS, combined with customer assessment, provide the documented processes for meeting the context requirements. The system must have adequate core programs to meet the needs of internal and external customers. The system must also have the relevant compliance programs to cover the hazards of its operations specifically.

Clause 4 also mentions expectations of workers and other interested parties placing a duty on the organization to determine expectations and needs. The reporting system in EOS protects anonymity and provides various avenues of reporting that overcome typical reporting barriers and allow for workers to voice concerns and communicate expectations. This is also an important component to the overall strategic communication strategy required by ISO 45001.

Clause 5 covers management's overall responsibilities, sets a style for policy, and requires worker participation. Strategic planning begins with the organization as a whole and then is mirrored in the leadership sections of the safety operations plan in EOS to meet the leadership and commitment section of clause 5. Organizational strategic plans that set a mission, guiding values, goals and objectives, communicate accountability, support, and integration of safety into core activities and facilitate a positive culture systematically. Conducting formal inventories, implementing a three-tiered hazard recognition program, creating an employee reporting system, continually reviewing management initiatives and investigational findings for effectiveness, empowering employees to increase competency levels and make decisions on the personal and program levels, measuring cultural criteria for performance, and tracking metrics based on the management duties, operational and cultural indicators, and final performance numbers, all combine to facilitate a positive organizational culture in which safety is a core component. ISO 45001 lists the following methods for which the organization's top management will demonstrate leadership and commitment:

1. Overall responsibility for prevention of work-related injury and illness
2. Provision of safe and healthy workplaces and activities

3. Incorporating the safety management system into the core business processes
4. Delivering necessary resources to the safety management system
5. Communicating importance and adhering to its own system policies
6. Ensuring that safety goals are achieved
7. Directing and supporting personnel to contribute to safety management system success
8. Practicing continual improvement
9. Supporting relevant management roles as they apply to respective areas of responsibility
10. Developing, leading, and promoting a culture that aligns with goals or the safety management system
11. Guaranteeing anonymity and non-retaliation for reporting safety-related issues and concerns
12. Allowing workforce inclusion at the consultative and participative levels
13. Supporting safety committee functioning (ISO 45001 5.1 2018)

Clause 5 also sets three overall duties for top management in regard to worker involvement. It must provide mechanisms, time, training, and resources necessary for consultation and participation. Management must also provide and give timely and clear understandable access and information about the safety management system. It must also remove or minimize barriers to participation. Involvement is mandated at the following levels for the relevant functions shown in Table 9.1.

TABLE 9.1 *Management Function and Involvement*

MANAGEMENT FUNCTION	WORKFORCE INVOLVEMENT LEVEL
Determining needs/expectations	Consultative
Establishing safety policy	Consultative
Determining roles, responsibilities, authority	Consultative
Determining how to meet legal and other requirements	Consultative
Establishing safety goals and objectives	Consultative
Planning for meeting goals and objectives	Consultative
Determining controls for contractors	Consultative
Determining what needs to be monitored, measured, and evaluated	Consultative
Establishing an audit program	Consultative

(Continued)

MANAGEMENT FUNCTION	WORKFORCE INVOLVEMENT LEVEL
Practicing continual improvement	Consultative
Creating mechanisms for workforce involvement	Participative
Hazard identification and risk assessment	Participative
Hazard and risk-abatement actions	Participative
Determining training needs/training	Participative
Creating strategic methods of communication	Participative
Formulating control measures in countermeasure implementation	Participative
Investigating incidents and forming countermeasures	Participative

(ISO 45001, Clause 5, 2018).

In employee-oriented safety, development of workforce-level associates to the next level of competency is an indication of program maturity. Consultation and actual participation and then allowing actual decisions to be made within organizational values are central to competency development. Numerous methods of consultation and participation can be implemented.

Hazard identification is a key element of clause 6. Hazard recognition is a function of all core management initiatives at some level. Hazards are not identified in one specific program. The three-tiered hazard-recognition program uncovers and empowers workforce associates to correct level 1 hazards, or hazards that match competency level. The reporting initiative uncovers hazards as concerns, questions, or reports of incidents. A thorough causal analysis in any investigation will often find combinations of causal factors that can be addressed specific to that incident and in similar activities or situations. An auditing component of an investigation program has the goal of proactive alleviation of substandard conditions. Behavior-based safety uncovers hazards associated with human behavior. The human factors program practices hazard and risk identification through early intervention concepts in design and modification and with muscular skeletal disorders. Even emergency response begins with an inventory and assessment of risk for hazards, threats, and vulnerabilities. The manners in which an organization accomplishes the first core duty of safety, identifying hazards, threats, and vulnerabilities, is across the board in many initiatives. ISO 45001 requires an "ongoing and proactive" program (ISO 45001 2018, 6.1.2.1). Hazard recognition must take into account how the work is organized and performed; social factors such as workload, equipment, materials, physical and environmental conditions; human factors; past organizational experiences; potential emergency situations; contractors; ground guests;

surrounding communities; organizational changes; modifications or new introductions to activities; changes in knowledge about existing hazards; and even hazards, threats, or vulnerabilities presented by situations or activities not under the control of the organization but with potential to impact safety and health of the workplace itself (ISO 45001 2018. 6.1.2.2).

Clause 6 describes EOS in that customers and situations are identified and the scope of the system is based on this initial concept and inventory of hazards, threats, and vulnerabilities. But the real key is countering the hazards, threats, and vulnerabilities in a manner that is documented through implementation and assessed for effectiveness. Clause 6 requires the organization to consider best practices, technological options, and financial, operational, and business requirements (ISO 45001 2018, 6.1.4). It is important to note that financial cost is not a basis for selecting safety controls and options. In case precedence, established in *Continental Oil Company v. Occupational Safety and Health Review Commission*, only an undue financial hardship, interpreted to severely threaten the businesses existence, justifies not upgrading controls to more reliable and technologically superior methods (630 F.2d 446, 1980.

The general planning portion of clause 6 requires standard practices when it comes to setting goals and objectives. Of course, workforce consultation is to be involved, and this can be an ongoing, continual process. Goals and objectives must be consistent with the established mission and guiding values. This is truly organizational ethics and is an indication of ethical conduct organizationally. Goals must also be measurable, demonstrable, or observable and take into account legal and ethical requirements, results of organizational experience, and assessments of hazards, threats, and vulnerabilities. These must also be communicated, monitored for performance, and reviewed and updated (ISO 45001 2018, 6.2.1). This process is basic management and has been described from the beginning of this system as the basis for functional management.

Clause 6 also provides a basic structure for strategic project planning that is standard. ISO 45001 requires that when planning for objectives, the plan must disclose what actions will be completed, the required resources needed, the responsible persons involved, projected dates of completion, how the results will be measured, and how the plan will be implemented into the business's processes (ISO 45001 2018, 6.2.2). Figure 9.6 displays an executive summary for such a project proposal from chapter 2. Notice that the executive summary contains the required elements and serves as a project proposal summary that can be followed by a more detailed report. Follow-up dates are projected dates of completion or stage completion.

Manufacturing LLC

Executive Summary

IC Number: _____   Date of Report: ____/____/_____

Approved By: _____   Facility: _____

To: Senior Management Team

Re: Installation of slip resistant rungs on facility permanent ladders

<u>Goal:</u> Reduce work comp expenditures for Slips/Falls on permanent ladders.

<u>Goal #1 from 2017 SOP:</u> XYZ Corporation will reduce slips on permanent ladder surfaces by 50% as compared and projected from the experience of slips on permanent ladders in 2018 at the XXville Facility.

<u>Justification:</u> Mission statement commits the organization to providing necessary resources for a safe and healthful work environment. In 2018 slips on permanent ladders were causal in 50 % of all slips, trips, or falls, totaling just over 18,000 dollars in loss. Projected cost of installation with outside resources is $9,000.00. Project break-even point is 6 months.

<u>Measure for Success:</u> After installation is complete: Number of STF's involving permanent ladders at XXville Facility for 1 year.

<u>Summary:</u> In 2017 Slips, Trips, Falls accounted for $xxxxx.xx in expenditures. STF's are #1 in frequency and expenditure, justifying the targeting of this condition. Of these STF's , 6 occurred due to non-compliant surfacing on fixed industrial ladders. Risk rating: High

<u>Project:</u> Purchase Slip-Not Brand rung covers.
Fabrication Section of Maintenance will install by welding:
North end of plant by XX/XX/XXXX   Responsibility: XXXXXXX Maint. Manager
South end of plant by XX/XX/XXXX Responsibility: XXXXXXX Maint. Manager

Resources Required	Resources On-Hand	Needed Resources/Cost	Strategy

Break- even Point: Experience Cost – Investment/time =
Follow-up: **Responsible Auditor:**

FIGURE 9.6 Executive Summary of Project Proposal

This suggested format goes beyond basic ISO 45001 requirements under clause 6.2.2 but is central in meeting other elements as it shows justification tied to strategic objectives and goals.

Clause 7 is about support and addresses the primary responsibility of upper management toward occupational safety, which is to provide the necessary resources. This is a critical measure of culture in the EOS system. Management commitment is measured by the allocation of resources from proposed safety projects that have been implemented. This begins with the chain of developing competency so that workers can be empowered. ISO 45001 requires the organization to determine necessary competency levels, ensure worker competency through education, training, and experience, and track this process (ISO 45001 2018, 7.2). Each initiative under EOS establishes competency and documents progression that includes successful experience with the goal of developing workforce-level associates to take even administrative responsibilities. Under EOS, competency is accounted for in each initiative due to workload and specificity in regard to the dynamics and knowledge of the individual initiative.

Clause 7 quickly turns toward strategic communications. Workforce-level associates must be made aware of or be routinely updated and communicated with regarding safety goals and objectives, safety policy, effectiveness of the overall safety system, implications of not conforming to safety system requirements, incidents and investigational outcomes, hazards and corrective actions relevant to the workers, and their right to work safe and stop work (ISO 45001 2018, 7.3). In the EOS system, workforce-level associates increase involvement to the empowerment level and are central in these overall functions.

The required communication plan must identify internal and external communication methods and identify what will be communicated, when, to whom, and how. These communications include internal communications to various levels, communications with contractors and visitors, and with the community. They must also consider aspects of diversity such as language and literacy disabilities (ISO 45001 2018, 7.4.1).

Communication plans can get very lengthy and involved. Communication is central in safety management and begins with reporting initiatives but includes training, safety committees, signage, educational postings, and others. Table 9.2 suggests a possible method of documenting the communications plan. The incident control log serves as the venue for documenting all safety-related activity, which would include any specific meetings or training deliveries.

TABLE 9.2 *Communications Plan Documentation*

COMMUNICATION	METHOD/DESCRIPTION	CUSTOMER	SCHEDULE
Safety orientation	Two-hour safety orientation	New hires/new temp workers	As needed
Visitor safety introduction	30-minute safety introduction video	Grounds visitor	As needed
Contractor safety meeting	Meeting with contractor to discuss safety system requirements	Contractors for new project	As needed
Annual community safety day	Half-day safety demonstration and facility tour/ annual town meeting-style briefing	Community/ associate families	Annual
Safety committee at large	Cross-functional representation from all safety initiative committees G&O and annual safety performance set	Internal associates	Quarterly; annual performance review and strategic plan

From these examples, the format is simple and documents the necessary information. The plan is not complete and typically would include the associate reporting methods such as the open door policy, safety concern communication, and all committees associated with the initiative such as the human factors committees, the system safety committees, or ad hoc problem solving committees. It may even contain a policy on email usage.

Clause 8 is operational and covers hazard abatement strategies, management of change, procurement, emergency planning, performance evaluation, auditing, and management review. ISO 45001 follows the direct ANSI abatement strategy for hierarchy of controls, which begins with elimination, continues with substitution, engineering controls, administrative controls, and personal protection (ISO 45001 2018, 8.1.2). In EOS, formulating countermeasures relies on two concurrent strategies where the psychological side of safety controls are enacted simultaneously to injury-prevention strategies that mirror the ANSI strategy and the ISO hierarchy of controls. The key difference is that psychological controls such as procedures, rules, education, training, signage, and awareness are implemented concurrently with any elimination or substitution goals and with specific engineering controls.

Management of change for the EOS system is covered as a system safety analysis policy. Mirroring the ISO 45001 requires any new workstation, process, or task introduced to the environment or any change to an existing workstation, process, or task to be reviewed in stages for early identification of problems and to ensure needs for competency development. According to ISO 45001, a good initiative for managing change considers workplace location, work organization, conditions, equipment, workforce, changes to legal and ethical requirements, new knowledge, and technology (ISO 45001 2018, 8.1.3).

Clause 8 also incorporates requirements for procurement of contractors. Many organizations use a process of prequalification and must assess contractors for meeting the safety management systems policies. The Multi-Employer Citation Policy (MECP) typically serves as a basis for overseeing contractors for safety operations but also serves as a starting point that mandates assessment of the contractor's safety history and present status in regard to the amount of supervisory authority that must be exhibited (OSHA 1999, CPL 2-0.124). ISO 45001 is concerned with identifying hazards and formulating controls for such hazards. This is also a central concern for meeting "reasonable care" as a controlling employer under the MECP. To meet "reasonable care," the controlling employer must conduct periodic inspections, implement an effective system for promptly communicating and correcting hazards, enforce compliance with a graduated system of enforcement, and follow up inspections (OSHA 1999, CPL 2-0.124). There are factors that influence the three duties for reasonable care and specifically determine the frequency of inspections. These include the scale of the project; the nature and pace of the work; the nature of the hazards to include their frequency, longevity, and severity; and the history of the contractor's safety performance, safety practices, and level of contractor expertise. More frequent inspections would be due for situations where the contractor had a history of noncompliance, no safety policy, no experience with this particular contractor, and other factors such as whether the contractor has a full-time safety professional or fails to have a dedicated safety manager (OSHA 1999, CPL 2-0.124).

Clause 8 also sets requirements for emergency preparedness and response. ISO 45001 sets seven elements:

1. Establish a planned response to emergency situations
2. Provide first aid
3. Train for a planned response, periodic tests, and exercises for response capability
4. Evaluate performance and review and revise plans after tests, exercises, and occurrences
5. Communicate duties and responsibilities to all workers
6. Communicate relevant information to visitors, contractors, community, and responding agencies
7. Consider the needs and capabilities of relevant parties, such as community responders (ISO 45001 2018, 8.2).

The emergency response plan outlined in the EOS system meets all seven elements of ISO 45001.

Clause 9 concentrates on monitoring, measurement, analysis and performance evaluation. Critical elements consist of evaluating compliance to legal requirements, establishing an audit program, and management review of the overall system and of individual initiatives. In the EOS system, all initiatives, core and compliance, have set metrics based on the Mathis structure of measures, as covered in chapter 2. The technique is also extended to measuring culture based on five criteria: participation, commitment, compliance, perception, and competency. Furthermore, these are divided into management level and workforce associate levels. This more than meets clause 9 requirements and is also used to prepare annual overall performance reviews of the system and to evaluate future direction for the system's strategic plans. These are all eluded too in previous clauses as well.

The internal audit program at the compliance level identifies areas of noncompliance to legal requirements and to company policy. All compliance initiatives have compliance-level audits to assess organization adherence, serve as basis for proactive formulation of controls, and document history of compliance based on relevant organizational needs. ISO 45001 outlines the elements of a proper internal audit program. It requires the following tasks:

1. Plan, implement, and maintain an audit program describing frequency of inspection, methods, responsibilities, reporting, correction process, and documented history
2. Set criteria and scope for each audit
3. Select objective auditors
4. Communicate findings to relevant managers, workers, representatives, and other interested parties
5. Take corrective actions
6. Document the history of findings and corrective countermeasures implemented (ISO 45001 2018, 9.2.2).

These are basic elements of any audit program, and documenting experience is critical for assessing culture. In the EOS system, management commitment is mainly measured from the percentage of countermeasures implemented. Implementation means that management has approved and allocated the necessary resources for meeting deficiencies.

Management review is a necessary component of continual improvement. In the EOS system, safety management reviews all initiatives as well as the overall program annually and, in some cases, such as when an emergency occurs, reviews actions as needed. Top management must review the overall performance of the safety program at least annually and based on the annual review of the overall program. A critical leading metric for this review is the measure of culture at the management and workforce levels.

Originating from review activity is the concept of continual improvement, as covered in clause 10 of ISO 45001. In the EOS system, all initiatives are assessed for effectiveness, which is management review, which also identifies weak areas where the initiative can be improved to increase overall safety effectiveness. ISO 45001 requires that the "organization continually improve the suitability, adequacy, and effectiveness" of the overall system (ISO45001 2018, 10.3). Continual improvement is accomplished through five criteria:

1. Enhancing system performance
2. Establishing a culture that supports the safety management system
3. Encouraging the participation of works for continual improvement of the system
4. Openly communicating relevant results of continual improvement
5. Documenting organizational experiences as evidence of continual improvement (ISO45001 2018, 10.3).

CONCLUSION

The management system established in this text as employee-oriented safety (EOS) originated from the community-oriented policing philosophy established for public safety management. It meets ISO 45001 or any other safety management system standard and provides some specific methods for which system requirements can be met. All parts of a system intermingle with one another and provide sustainable management practices that survive associate turnover and business interruptions. Central to any system is its assessment for the needs of the system and overall management philosophy. EOS relies on viewing the safety department or function as an equal component to the other cross functional sections of the organization and to the relevant external components of the organization. Internal and external parties are viewed as customers who have needs, and to fulfill these needs, two-way communication and partnership is established based on equality, trust, transparency, and involvement that develops into empowerment.

REFERENCES

International Organization for Standards. 2018. *International Standard ISO 45001: Occupational Health and Safety Management Systems—Requirements with Guidance for Use*, 1st ed. Vernier, Geneva: Author.

Occupational Safety and Health Administration. 1999. *Multi-Employer Citation Policy* (CPL 2-0.124). Washington DC: Author.

US Department of Labor. Multi-Employer Citation Policy. https://www.osha.gov/enforcement/directives/cpl-02-00-124

CPSIA information can be obtained
at www.ICGtesting.com
Printed in the USA
FSHW011649181119
64242FS

9 781516 527823